T0137110

Springer Theses

Recognizing Outstanding Ph.D. Research

Aims and Scope

The series "Springer Theses" brings together a selection of the very best Ph.D. theses from around the world and across the physical sciences. Nominated and endorsed by two recognized specialists, each published volume has been selected for its scientific excellence and the high impact of its contents for the pertinent field of research. For greater accessibility to non-specialists, the published versions include an extended introduction, as well as a foreword by the student's supervisor explaining the special relevance of the work for the field. As a whole, the series will provide a valuable resource both for newcomers to the research fields described, and for other scientists seeking detailed background information on special questions. Finally, it provides an accredited documentation of the valuable contributions made by today's younger generation of scientists.

Theses are accepted into the series by invited nomination only and must fulfill all of the following criteria

- They must be written in good English.
- The topic should fall within the confines of Chemistry, Physics, Earth Sciences, Engineering and related interdisciplinary fields such as Materials, Nanoscience, Chemical Engineering, Complex Systems and Biophysics.
- The work reported in the thesis must represent a significant scientific advance.
- If the thesis includes previously published material, permission to reproduce this must be gained from the respective copyright holder.
- They must have been examined and passed during the 12 months prior to nomination.
- Each thesis should include a foreword by the supervisor outlining the significance of its content.
- The theses should have a clearly defined structure including an introduction accessible to scientists not expert in that particular field.

More information about this series at http://www.springer.com/series/8790

Riccardo Freccero

Study of New Ternary Rare-Earth Intermetallic Germanides with Polar Covalent Bonding

Beyond the Zintl Picture

Doctoral Thesis accepted by
University of Genoa, Genoa, Italy

 Springer

Author
Dr. Riccardo Freccero
Chemical Metals Science
Max Planck Institute for Chemical Physics
of Solids
Dresden, Germany

Supervisors
Dr. Frank R. Wagner
Chemical Metals Science
Max Planck Institute for Chemical Physics
of Solids
Dresden, Germany

Prof. Adriana Saccone
Department of Chemistry and Industrial
Chemistry
University of Genoa
Genoa, Italy

ISSN 2190-5053   ISSN 2190-5061   (electronic)
Springer Theses
ISBN 978-3-030-58994-3   ISBN 978-3-030-58992-9   (eBook)
https://doi.org/10.1007/978-3-030-58992-9

This Springer imprint is published by the registered company Springer Nature Switzerland AG
The registered company address is: Gewerbestrasse 11, 6330 Cham, Switzerland

"Curiosity?" asked Mr. Eastman

"Yes", I replied, "curiosity", which may or may not eventuate in something useful, is probably the outstanding characteristic of modern thinking.
It is not new.
It goes back to Galileo, Bacon and to Sir Isaac Newton.

Abraham Flexner

And the role of philosophy is precisely that of revealing to men the usefulness of the useless or, if you will, to teach them to distinguish between the two meanings of the word useful.

Pierre Hadot

Supervisors' Foreword

For chemists, polar intermetallic phases probably represent the most challenging crystalline compounds from a synthetical, structural and conceptual point of view. The composition, atomic arrangement and physical properties of the compounds of this family continue fascinating researchers around the world.

Compared to typical Zintl phases, the polar intermetallic compounds are characterized by a smaller electronegativity difference between the constituting electropositive and electronegative species, which results in a reduced charge transfer. This leads to unexpected chemical bonding scenarios where the cations–anions interactions are often described as polar covalent, instead of ionic. Polar intermetallics are characterized by an average number of electrons per atom that is intermediate between those of Hume-Rothery phases and those of Zintl phases. Unfortunately, being quite difficult to find a critical e/a range for the formation of polar intermetallics, the literature reports different relative phase distributions as a function of *e/a*. Such tentative determinations constitute nevertheless valuable attempts at finding predictive electron counting rules; however, more challenging experimental and theoretical efforts have to be performed in order to fully disclose the fascinating and complicated chemistry of polar intermetallic phases.

The intriguing, fully original and far from trivial work performed by the Ph.D. Riccardo Freccero on new polar intermetallic compounds is of considerable scientific interest and practical relevance. The Ph.D. work, performed in the framework of a cooperation between the Dipartimento di Chimica e Chimica Industriale of the Università degli Studi di Genova (Italy) and the Max-Planck-Institut für Chemische Physik fester Stoffe of Dresden (Germany), achieved the planned objectives by a combination of experimental and theoretical cutting-edge techniques.

The introductory Chap. 1 of the thesis provides a brief introduction to the Zintl and polar intermetallic phases that exhibit complex bonding scenarios from which arise unprecedented physical properties with promising applications in different fields such as superconductors, thermoelectrics, hydrogen storage and zero-thermal expansion materials.

Binary and ternary rare-earth germanides include a numerous family of polar intermetallics showing both rich structural chemistry and interesting properties. These polar intermetallics, often existing as continuous series throughout the lanthanide family, are indeed excellent candidates for the investigation of structure–property relationships. Therefore, the starting point of this Ph.D. thesis was the Ge-rich corner of the La–Mg–Ge phase diagram, which had been previously investigated by the Genova group, where the two germanium-rich polar intermetallic compounds La_4MgGe_{10-x} and La_2MgGe_6, exhibiting structural peculiarities such as Ge sites deficiency and Ge covalent fragments, were detected and characterized. The doctorate work attained the synthesis, the crystal structure characterization and the analysis of the chemical bonding, also using the most recent real space techniques, of different rare-earth series such as R_2MGe_6. Particular attention was also given to the role of the M metals, for instance, in the R_4MGe_{10-x} series (M = Li and Mg).

In Chap. 2, the author describes the two synthetic routes used in order to obtain the rare-earth ternary germanides investigated and also to evaluate the influence of the synthetic method on their stability and formation: direct synthesis through arc/induction melting or in resistance furnace, and metal flux synthesis in resistance furnace. A brief description of the characterization techniques used, LOM, SEM, EDXS, XRD (X-ray powder and single crystal diffraction), DTA, closes the chapter.

In Chap. 3, the author describes theories and computational methods adopted to obtain/predict energies and properties of many solid systems starting from the Density Functional Theory (DFT) and proceeding with the Quantum Theory of Atoms In Molecules (QTAIM) and the Electron Localizability Indicator (ELI). In particular, improved methods were employed to analyze the chemical bonding of the R_2MGe_6 intermetallics which allow to adequately characterize different bonding interactions like those of La–Ge and M–Ge. The chemical bonding analysis goes beyond current state-of-the-art quantum chemical techniques by proposing a new and sophisticated tool to uncover a chemically useful fine structure of the electron distributions related to polycentric bonding features not easy to extract otherwise.

Chapter 4 (on R_2MGe_6 compounds), Chap. 5 (on $R_2Pd_3Ge_5$ compounds), Chap. 6 (on $Lu_5Pd_4Ge_8$, $Lu_3Pd_4Ge_4$ and Yb_2PdGe_3 compounds) and Chap. 7 (on R_4MGe_{10-x} compounds, M = Li, Mg) describe the results obtained on these intermetallics series from different points of view: synthesis, crystal structure characterization (determination of crystallographic symmetry group–subgroup relations, structural chemical relationships among the various structure types investigated), precise DFT calculations (total energy, electronic structure, effective charges and charge transfer, chemical bonding), physical properties (measurement and interpretation of magnetism and electrical conductivity as well as Seebeck coefficient in selected compounds).

Chap. 8 is reserved for conclusions. Even though accurate chemical bonding investigations were not performed for all the compounds, the author proposed a valid generalized scheme, according which the investigated compounds should be treated as polar intermetallics where a partial charge transfer occurs. As a consequence, additional covalent interactions, even polycentric, take place beyond the Ge–Ge ones. Only with M = Li, Mg, the Ge–M interactions can be described as mainly ionic. With the other M metals, the description of Ge–M interactions is more appropriately described as heteropolar. Moreover, this study highlights that the M–R two-centre polar bonds play a crucial role when M is a transition metal.

The results achieved can constitute a step forward in a better comprehension of composition–structure–properties relationships in the field of polar intermetallics, and we are confident that future research work on this subject will profit by this study.

Genoa, Italy
Dresden, Germany
September 2020

Prof. Adriana Saccone
Dr. Frank R. Wagner

Abstract

The syntheses, structural characterizations and chemical bonding analyses for several ternary R–M–Ge (R = rare-earth metal; M = another metal) intermetallic compounds are reported. Each chapter of the thesis, the introductory ones apart, is dedicated to the obtained achievements for a specific series of investigated compounds, which are: R_2MGe_6 (M = Li, Mg, Al, Cu, Zn, Pd, Ag), R_4MGe_{10-x} (M = Li, Mg), $R_2Pd_3Ge_5$, $Lu_5Pd_4Ge_8$, $Lu_3Pd_4Ge_4$ and Yb_2PdGe_3.

Preparation techniques included both traditional and innovative methods, like the metal flux synthesis, which turned out to be crucial for crystal growth and stabilization of some metastable compounds.

Accurate crystal structure determinations were performed on the basis of both single crystal and powder diffraction data. In the case of R_2LiGe_6, R_4MGe_{10-x} and $Lu_5Pd_4Ge_8$, the presence of non-merohedrally twinned crystals was successfully faced. The obtained structures for the R_2MGe_6, $R_2Pd_3Ge_5$, $Lu_5Pd_4Ge_8$ and Yb_2PdGe_3 were concisely described and rationalized according to the group–subgroup formalism. These results, combined with total energy calculations, allow presenting a correct distribution of structure modifications among the large family of the R_2MGe_6 compounds, leading to a deep revision of many controversial literature data.

The knowledge of the correct structural models was the essential starting point to perform accurate and reliable chemical bonding investigations, mainly focusing on the far from trivial interactions between the Ge-polyanionic networks and the surrounding metals, revealing in all cases strong deviation from the Zintl description. In this framework, a comparative chemical bonding analysis for La_2MGe_6 (M = Li, Mg, Al, Zn, Cu, Ag, Pd) and Y_2PdGe_6 germanides was performed by means of cutting-edge position-space quantum chemical techniques based on QTAIM, ELI-D and their basins intersection. The accurate description of the bonding scenario required also the proposal of two new approaches: the penultimate shell correction (PSC0) and the ELI-D fine structure analysis based on its relative Laplacian. Hence, the Ge–La/Y and Ge–M ($M \neq$ Li, Mg) bonding were described as polar covalent. The Li- and Mg-containing phases were described as germanolanthantes $M[La_2Ge_6]$. Finally, thanks to these tools, a consistent picture for La/Y–M polar bonds was also presented.

Publications Related to this Thesis

Many of the obtained results were published in five papers. Parts of them are reported/reproduced in this thesis with the editors' permissions, when necessary. More details are listed in the following:

1. R. Freccero, P. Solokha, S. De Negri, A. Saccone, Yu. Grin, F.R. Wagner, "Polar-covalent bonding beyond the Zintl picture in intermetallic rare-earth germanides" *Chem. Eur. J.*, 2019, *25*, 6600–6612 (https://doi.org/10.1002/chem.201900510).
   Reported in Sects. 3.4, 3.5, 4.2, 4.6 and 4.7.

2. R. Freccero, P. Solokha, D.M. Proserpio, A. Saccone, S. De Negri, "A new glance on R_2MGe_6 (R = rare earth metal, M = another metal) compounds. An experimental and theoretical study of R_2PdGe_6 germanides" *Dalton Trans.*, **2017**, *46*, 14021–14033 (https://doi.org/10.1039/c7dt02686b).
   Reported in Sects. 4.1–4.5 and 4.7

3. P. Solokha, R. Freccero, S. De Negri, D.M. Proserpio, A. Saccone, "The $R_2Pd_3Ge_5$ (R = La–Nd, Sm) germanides: synthesis, crystal structure and symmetry reduction" *Struct. Chem.*, **2016**, *27*, 1693–1701 (https://doi.org/10.1007/s11224-016-0812-z).
   Reported in Sects. 5.1 and 5.3

4. R. Freccero, S.H. Choi, P. Solokha, S. De Negri, T. Takeuchi, S. Hirai, P. Mele, A. Saccone, "Synthesis, crystal structure and physical properties of $Yb_2Pd_3Ge_5$" *J. Alloys Compd.*, **2019**, *783*, 601–607 (https://doi.org/10.1016/j.jallcom.2018.12.306).
   Reported in Sects. 5.2 and 5.3

5. R. Freccero, P. Solokha, D.M. Proserpio, A. Saccone, S. De Negri, "$Lu_5Pd_4Ge_8$ and $Lu_3Pd_4Ge_4$: Two More Germanides among Polar Intermetallics" *Crystals*, **2018**, *8*, 205 (https://doi.org/10.3390/cryst8050205).
   Reported in Sect. 6.1

Acknowledgements

The Ph.D. time has ended and, for the first time after about twenty-one years, I'm not anymore a "conventional" student. If I turn back and look at the path I've travelled I see once again, with wonder and deep feeling, the huge amount of people I should thank for their teaching, support, help, patience and love. Alone, nothing would have been possible! Although it could sound quite weird, I guess that a very nice word that applies to all the people I would like to thank, both from the scientific and human point of view, is *useless*. I will try to make this statement slightly clearer by quoting a very short sentence: "*Useless is everything that helps us become better.*" (Nuccio Ordine, *The usefulness of the useless*).

For the "scientific part", it is now perhaps clear to the reader why I placed at the beginning of this manuscript a short citation taken from "*The usefulness of useless knowledge*" written by the renowned American scientist and pedagogue, Abraham Flexner, in 1937. Firstly, I would like to begin by thanking all scientists that, in these last eight years, taught me with expertise and passion, the love for the *useless* science without feeling embarrassed when, after being asked the question "Why do you investigate intermetallic chemistry? What is it useful for?", I replied "Because it is awesome: it considerably expands our chemical knowledge!". Among them, I would like to thank first my supervisors, Prof. Adriana Saccone and Dr. Frank R. Wagner. Prof. Saccone, starting from the second year of the bachelor when she introduced me to the fascinating world of the chemistry of the elements, was a constant presence in each of my steps in the borderless world of research and solid-state chemistry. Dr. Wagner is the one that introduced and guided me into the fields of computational and theoretical chemistry, which were unexplored "lands" for me. Together, we faced the issue of investigating chemical bonding for selected compounds: it has been a long and interesting adventure! Prof. Yuri Grin made it possible by hosting me in Dresden, at the Max-Planck-Institute for Chemical Physics of Solids, both during my master and my Ph.D., starting a fruitful and enriching collaboration: thanks! Heartfelt thanks to Serena. Although unofficially, she has been a constant and essential presence in my entire short scientific career. She taught me a huge amount of notions in many aspects concerning the field of research: of course, chemistry but also how to write a paper, how to submit it, how a conference is usually structured, and so on and so forth... But I'm particularly thankful to her for the friendship, for our long, sometimes *useless* and wonderfully nice chats. My esteem for her lead, as a natural consequence, to chose her as one of my maids of honour: thank you! She shared the far from trivial duty to support and bear me, every day, with Pavlo. Again, it is hard to recall how many things he taught me, starting from the first steps "within crystal structures" well before the Ph. D. Thank you for all the stimulating discussions and for your never-ending availability to help me and to share a resting coffee break, introduced by the sweet sound of the "coffee bell". Additional special thanks go to Simona, Enrico, Federica and Riccardo. Each of you has taught me something: I'll avoid listing it here for conciseness but it is stored in my mind. Anyway, apart from teaching, I would really like to say (once again) thank you for all the *useless* moments we have shared during my Ph.D. These I also do not list, but they are all safely stored in my heart.

Coming back to Dresden, I would also like to thank my officemates Ana, Olga, Karsten, and Julia, Sévère and Oksana for making my stay in Germany enjoyable also from the human point of view: thank you for all the Italian coffee breaks in the winter garden and for the friendship. It was also worth coming to Dresden just to meet you!

Cristina. It is hard to find the appropriate words to thank you. The period of the Ph.D. is studded with unforgettable shared everyday life and special moments, achieved goals and new beginnings. Thank you for your constant and sensitive presence: I'm not able to imagine the Ph.D. without you: impossible! But I would also, and particularly, say thanks to you for allowing me to be at the same time a constant presence in your intense life, which has passed through the master studies and the new job. I'm incredibly thankful to you for this everyday common path, which culminated with our wedding. I've done nothing to deserve so much: thanks!

I'm grateful to my mum, Marcella, and my dad, Pino. Thank you for growing me with limitless love and patience, which you needed a lot of. Thank you for teaching and showing me "the art of life", which I could say is "the *useless* art to love", in a very concrete way which includes also, of course, facing with perseverance and strength difficulties. To my young brother, Alberto: thank you! Thanks for your incredible esteem and love. You are a special person and I'm grateful to have shared the same roof and grown with you for so many years! Thank you also to my grandparents, Rita, Adelina, Nicco and "Rossi" and to all my family. I would also thank my "enlarged" family in particular, my in-laws Patrizia and Antonio, and my brother-in-law Andrea for their warm hospitality and increasing affection.

Since I have mentioned the wedding, I would like to thank some friends that have played a particular role accompanying us to that day: Danilo, Antonio, Federica and Armanda. To the guys of the Arco group: thank you! It is not possible to thank all friends (they are really many!) and people that had a decisive role in my growth and education. I can say that I've been really lucky and that I'm of course aware of their importance. For shortness, I've tried to focus, sometimes without success, in the three years of the Ph.D.

I really hope that some people that I've mentioned above and also those that I've not, enjoyed, at least once, my *useless* company. Thanks again.

Riccardo Freccero

Contents

Chapter 1
Introduction

The solid state is one of the fundamental states of the matter and probably the one most heavily utilized by human civilization in any period of its development, starting from the stone age, through the copper, brass and iron ages, until today. In the past the technological relevance of metals, like copper and iron, and alloys, like brass, was already clear. Nowadays, many modern technologies either as constructional or as data storage, are based on solids and, in particular, on metallic materials. That is why many research efforts are focused on the preparation of new solid materials. Anyway, the reasons why certain compositions result with specific structures are still far to be clearly understood and would be of great importance to plan syntheses of compounds with selected properties. Therefore, from a chemical point of view, a deeper knowledge of the chemical bonding for such compounds would help to rationalize the just mentioned composition-structure–property relationships.

According to the type of chemical bonding, the crystalline solids are generally classified into three principal groups: metallic, ionic and covalent. An useful scheme including all of them is the van Arkel-Ketelaar bonding triangle [1, 2] shown in Fig. 1.1.

The average electronegativity $\overline{\chi}$ of the constituents is reported on the horizontal axis (triangle base) from the most electropositive Caesium (Cs) to the most electronegative Fluorine (F); the electronegativity difference between the constituting elements is reported on the vertical axis (triangle height), so that the top vertex corresponds to Caesium fluoride CsF (maximum $\Delta\chi$). Within this representation the main classes of crystalline solids are located at the three corners:

- Left corner hosts classical metallic compounds: all the constituting elements are electropositive with highly delocalized valence electrons.
- Right corner corresponds to covalent compounds: valence electrons tend to be localized between atoms.

R. Freccero, *Study of New Ternary Rare-Earth Intermetallic Germanides
with Polar Covalent Bonding*, Springer Theses,
https://doi.org/10.1007/978-3-030-58992-9_1

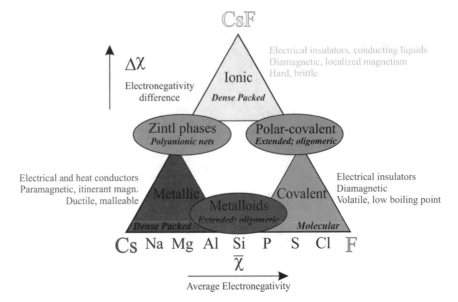

Fig. 1.1 The van Arkel-Ketelaar triangle. The main classes of solid compounds together with their physical properties are highlighted. Adapted with permission from [3, 4]

- Top corner hosts ionic compounds: valence electrons are localized around the more electronegative elements.

Each of these classes of materials is characterized by peculiar properties which arise from the different kind of chemical bonding: some of these properties are listed near the corresponding region (Fig. 1.1). For example, electrical conductivity occurs with metals, while covalent and ionic compounds are electrical insulators; on the other hand, ionic compounds are conducting in the liquid state. From the point of view of mechanical properties metals are mainly ductile, while ionic compounds are brittle.

More complicated chemical bonding scenarios are realized in intermediate regions of the triangle, leading to different classes of compounds. From these unusual bonding regimes, unprecedented or enhanced properties can arise, and this is often a strong motivation for their investigation.

Among these "non classical" materials, located somewhere between "pure" metallic systems and covalent and/or ionic compounds, Zintl and polar intermetallic phases are of great interest.

An intermetallic compound, as defined by Shulze [5], is a solid phase containing two or more metallic elements, with optionally one or more non-metals, the crystal structure of which differs from that of the constituents.

A common classification of intermetallic compounds is based on the main factor governing their crystal structure. For Laves and Hägg phases it is the geometrical/size factor, for Hume-Rothery phases it is the valence electron concentration per formula unit (*VEC f.u.*) and for Zintl phases [6, 7] it is the electrochemical factor (electronegativity difference). The binary Zintl phases are usually composed by an electropositive metal (*e.g.* belonging to alkali, alkaline earth metals), often called "active metal", and a more electronegative element around the so called "Zintl line", placed between the groups 13 and 14 of the periodic table.

A simple but powerful rule to rationalize the structures of Zintl phases is the Zintl-Klemm formalism. The essence of this approach is the formal charge transfer of valence electrons from the electropositive to the electronegative atoms which induce the latter to establish covalent bonds in order to join the octet stable electronic configuration. Klemm generalized this powerful idea by introducing the *pseudoatom concept* for the electronegative components, which means these elements would exhibit a structure based on nearest-neighbor atomic connectivity characteristic of the isovalent element (see the KGe example below). In this way different kind of chemically interesting polyanions, charge balanced by electropositive species, may form. Ferro and Saccone [7] proposed to separate Zintl phases into three groups depending on the nature of the polyanionic network:

(a) *Zintl phases with delocalized bonding*, corresponding to isolated clusters, mainly of the elements of the 13th and 14th groups, whose structure can be generally rationalized with the Wade–Mingos rules [8, 9].
(b) *Zintl phases with interconnected (homo- and hetero-atomic) clusters*, mainly given by alkali metals and 13th groups elements. Generally, in these compounds two types of bonding are combined: delocalized within the clusters (endo-bonds) and localized ($2c$, $2e$ exo-bonds) between them.
(c) *Zintl phases with localized bonding*. Compounds containing only $2c$, $2e$ (two centres and two electrons) bonds among the p–elements. The anionic components behaviour can be rationalized applying the octet rule and thus they generally follow the Klemm pseudoatom concept.

A typical example of polyanionic Zintl compound, belonging to the third group (c) among those listed above, is the well-known potassium germanide (KGe [10]). According to Zintl-Klemm rules, K formally donates its $4s$ valence electron to Ge which then behaves as a *pseudo-pnictogen* atom, forming three covalent bonds with four neighbouring Ge$^-$ ions, constructing isolated anionic Ge_4^{4-} clusters (Fig. 1.2, to the left). The latter, are isostructural with the naked P_4 molecular clusters found in the white phosphorus unit cell (Fig. 1.2, to the right).

Potassium cations occupy interstitial sites so that electroneutrality is fulfilled.

In order to rationalize all the aforementioned Zintl phases, the partial *VEC* is a powerful tool [7, 11]. For a C_mA_n compound (*C*: cation; *A*: anion) the anionic partial valence-electron concentration (VEC_A) is:

$$VEC_A = \frac{m \cdot e_C + n \cdot e_A}{n} \qquad (1.1)$$

where:
m, n = composition indexes.
e_C, e_A = valence electrons of the elements *C* and *A*.

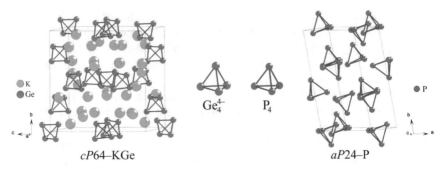

Fig. 1.2 Unit cell representations for KGe (to the left) and white P (to the right). Covalent Ge–Ge and P–P bonds are represented with red sticks. The isostructural tetrahedral molecular clusters are clearly visible

This formalism was first applied to typical ionic compounds, where the cations transfer the exact number of electrons to complete the anions valence shells. In this case $VEC_A = 8$. Generally, for (polyanionic) Zintl phases VEC_A is < 8. Compounds showing a $VEC_A > 8$ have been described by certain authors as polycationic Zintl phases [11], which are not treated in this thesis.

Once the VEC_A (<8) is obtained, it is possible to evaluate the average number of anion-anion homocontacts, $b(A–A)$, using the "8–VEC_A" rule, which is effectively the octet rule (often reported as 8–N rule). This general approach applied to KGe intermetallic give results in agreement with previous considerations based on the interatomic distances analysis:

$$VEC_{Ge} = \frac{1+4}{1} = 5 < 8 \Rightarrow \text{polyanionic}$$
$$b(Ge-Ge) = 8 - VEC_{Ge} = 3 \tag{1.2}$$

Hence, in spite of its simplicity, the Zintl-Klemm approach is successful in structural rationalization and can be extended also to some ternary, quaternary and multicomponent compounds.

Another classification of Zintl phases which also takes into account physical properties, was proposed by Nesper [12]. He defined three criteria:

(1) a well-defined relationship exists between their chemical (geometrical) and electronic structures (i.e., certain aspects of their structures satisfy electron counting rules);
(2) they are semiconductors (energy gaps less than 2 eV), or, at least, show increasing electrical conductivities with increasing temperature;
(3) they are mostly diamagnetic, but, if paramagnetic, they should not show temperature-independent (Pauli) paramagnetism, typical for metallic systems.

Fig. 1.3 Relative
distributions of intermetallic
phases in terms of *e/a* values
according to Corbett et al.
[15] (**a**) and Dronskowski
et al. [16] (**b**). Figure a is
reproduced with permission
from [15]

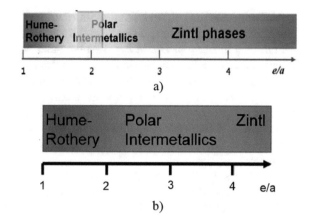

These criteria imply that Zintl phases (like valence compounds) have narrow homogeneity widths (i.e., they are mostly line compounds) and could display semimetallic behaviour.

Compared to typical Zintl phases, the polar intermetallic compounds are characterized by a smaller electronegativity difference ($\Delta\chi$) between the constituting electropositive and electronegative species which results in a reduced, and then partial, charge transfer. It happens when the combined elements are "getting closer" in the periodic table, e.g. the electronegative component progresses from the right to the left of the Zintl border. This lead to unexpected and unpredictable chemical bonding scenarios where the cations-anions interactions were often described as polar-covalent, instead of ionic; as a results, their structures typically do not follow the 8–N rule and the Nesper criteria being often metallic. Anyway, for some compounds, the Zintl-Klenm rule seems to be, at least formally, still valid and can be helpful to preliminary analyse the presence of polyanionic fragments. For instance, it is the case of the Ca_5Ge_3 intermetallic [13] where, on the basis of valence electron count, Ge_2 dumbbells and isolated Ge species can be guessed, leading to $\left(Ca^{2+}\right)_5\left[(1b)Ge^{3-}\right]_2\left[(0b)Ge^{4-}\right]$ ionic formulation, in very good agreement with Ge–Ge interatomic distances analysis. Nonetheless, electronic structure calculations revealed the presence of bonding covalent interactions between Ca and Ge, responsible for its conductivity. Some author called this kind of compounds "metallic Zintl phases" [14].

At this point, it is worth to note that there is not a specific critical $\Delta\chi$ value that separates typical Zintl phases from polar intermetallics. The total valence electron concentration per formula unit, which represents the average number of electrons per atom (*e/a*) [15], was proven to be an useful parameter to classify different kind of intermetallics.

The schemes in Fig. 1.3 [15, 16] show that polar intermetallics are characterized by *e/a* values intermediate between those typical for Hume-Rothery phases and those characteristic of Zintl phases.

Unfortunately, it is quite difficult to find a critical e/a range for the formation of polar intermetallics. In fact, different authors report different relative phase distributions as a function of e/a. In Fig. 1.3a the polar phases tend to form with an e/a values close to 2 ± 0.5 whereas in Fig. 1.3b their existence field extend till $e/a \sim 3.5$.

These two e/a scales constitute valuable attempts to find out working electron counting rules, with predictive purposes, analogously to what it's known for Zintl and Hume-Rothery phases. Nevertheless, such rules are still far to be definitely formulated and more challenging experimental and theoretical efforts have to be performed in order to disclose the fascinating and complicated polar intermetallic chemistry.

The interest for Zintl and polar intermetallics is not only related to their structural and chemical bonding peculiarities, but also to their intriguing physical properties. In fact, some of these phases have been proved to be promising for different applications, *i.e.* as superconductors, thermoelectrics, hydrogen storage and zero-thermal expansion materials [17].

Binary and ternary rare earth germanides constitute a numerous family of polar intermetallics, which was intensively studied because of their rich structural chemistry and interesting properties. Such compounds are excellent candidates for studies on structure – property relationships often existing as continuous series throughout the lanthanide family.

During the doctorate work the research efforts were put on the synthesis, crystal structure, property measurements and chemical bonding investigation on polar R-M-Ge germanides (R = rare earth metal, M = Li, Mg, Al, Cu, Zn, Pd, Ag) of different stoichiometries.

1.1 The Ge-Rich Corner of the La–Mg–Ge Phase Diagram: The Starting Point

During the investigation of the La–Mg–Ge phase diagram at 500 °C (Fig. 1.4a), carried out by De Negri et al. [18], two germanium rich polar intermetallic compounds (Fig. 1.4b) were detected and characterized: La_4MgGe_{10-x} and La_2MgGe_6. The former constitutes a new structure prototype whereas the latter, crystallizing in the $oS72$-$Ce_2(Ga_{0.1}Ge_{0.9})_7$ structure, belongs to the numerous R_2MGe_6 family (M = transition or main group metal). For M = Mg this compound was found to exist only for R = La [19].

After studying these two new polar intermetallic compounds, the interest for these phases grew up in the group, mostly in relation to their structural peculiarities such as Ge sites deficiency, Ge covalent fragments, etc....

The main purposes of the doctorate work were the syntheses (also with metal flux), crystal structure characterization/analysis and chemical bonding (also with the most recent real space techniques) investigation of different R_2MGe_6 with particular interest on the different role of the M metals (see Chap. 4). The existence of the

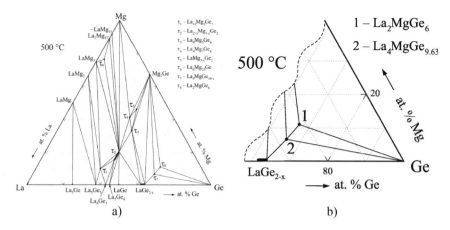

Fig. 1.4 a Isothermal section at 500 °C for La–Mg–Ge system. All ternary phases found are listed. **b** Ge rich corner of the phase diagram: the La_2MgGe_6 and La_4MgGe_{10-x} germanides were detected. Reproduced with permission from [18]

R_4MGe_{10-x} was also checked for M = Li and Mg (Chap. 7) along the lanthanide series. Results obtained during these syntheses often lead to the formation of some new interesting compounds which constituted new starting points for other experimental and theoretical investigations. In such a way, the structural/chemical variety of different R–M–Ge intermetallics, in particular with M = Pd (Chaps. 5 and 6), were described allowing also to deeply understand the main chemical interactions acting among the involved elements.

References

1. van Arkel AE (1956) Molecules and crystals in inorganic chemistry. Interscience, New York
2. Ketelaar JAA (1958) Chemical constitution; an introduction to the theory of the chemical bond, 2nd edn. Elsevier, Amsterdam
3. Miller GJ, Zhang Y, Wagner FR (2017) Chemical bonding in solids. In Richard D, Shinichi K, Andreas S (eds) Handbook of solid state chemistry, 1st edn. Wiley-VCH Verlag GmbH & Co KGaA. https://doi.org/10.1002/9783527691036.hsscvol5013
4. Miller GJ (2014) Chemical bonding and electronic structure of solids. Lecture given at TU Dresden
5. Schulze GER (1967) Metallphysik. Akademie-Verlag, Berlin
6. Zintl E, Dullenkopf W (1932) Phys Chem B16:183
7. Ferro R, Saccone A (2008) Intermetallic chemistry. Elsevier Pergamon Material Series
8. Wade K (1976) Adv Inorg Chem Radiochem 18:1
9. Mingos DMP (1977) Adv Organomet Chem 15:1
10. Busmann E (1960) Naturwissenschaften 47:82
11. Müller U (2006) Inorganic structural chemistry, 2 edn. Wiley
12. Nesper R (1990) Prog Solid State Chem 20:1
13. Mudring AV, Corbett J (2004) J Am Chem Soc 126:5277–5281
14. Miller GJ, Lee C-S, Choe W (2002) In: Meyer G, Naumann D, Wesemann L (eds) Inorganic chemistry highlights. Wiley, Weinheim

15. Lin Q, Corbett J (2009) A chemical approach to the discovery of quasicrystals and their approximant crystals. In: Mingos DMP (ed) Structure and bonding. Springer-Verlang, Berlin, Heidelberg. https://doi.org/10.1007/430_2008_11
16. Steinberg S, Dronskowski R (2018) Crystals 8:225. https://doi.org/10.3390/cryst8050225
17. Fässler TF (2011) Zintl phases: principles and recent developments. Springer-Verlag, Berlin Heidelberg
18. De Negri S, Solokha P, Skrobańska M, Proserpio DM, Saccone A (2014) J Solid State Chem 218:184–195. https://doi.org/10.1016/j.jssc.2014.06.036
19. Freccero R, Solokha P, De Negri S, Saccone A, unpublished results

Chapter 2
Experimental Methods

2.1 Sample Synthesis

In order to evaluate the influence of the synthetic method on the stability and formation of the investigated rare earth ternary germanides, different synthetic routes and
methods were followed:

- Direct synthesis through arc/induction melting or in resistance furnace
- Metal flux synthesis in resistance furnace

In all cases the starting elements were from commercial sources with nominal
purities always higher than 99.9 mass %. More details are given in Table 2.1

All metals, before usage, were mechanically freed from surface oxide layers.
Samples containing particularly oxidizable elements, like Li, were weighed within
an Ar filled glovebox (H_2O and O_2 levels < 0.1 ppm; MB 10 G, MBraun). Except the
case of arc melted samples, it was necessary to choose appropriate crucible materials
depending on the involved elements which should not react with the containers.
Alumina (Al_2O_3) and tantalum (Ta) were selected for this purpose. To ensure their
cleanness, the former were sonicated in aqua regia and then heated for about twelve
hours at 1200 °C in resistance furnace; the latter were sonicated within acetone for
a few minutes and subsequently heated under Ar flow in induction furnace. For this
work, the access to custom-made arc-sealed Ta crucibles was crucial for the synthesis
involving metals with high vapour pressure (*e.g.* Mg or Yb), giving the possibility
to prevent reactants losses. In the case of metal flux synthesis, Ta crucibles were
produced according to the chosen excess of the metal solvent. Aiming to prevent
high temperature oxidation of the samples and of Ta, induction melting was always
performed under Ar flow; when using the resistance furnace, the filled crucibles
were closed in evacuated quartz phials. During both direct and metal flux synthesis,
a continuous rotation, at about 100 rpm, of the phial was applied (Fig. 2.1). In this way

Table 2.1 Starting elements together with their sources and shape

Element	Source	Shape
Ge	MaTecK	chunk
R^*	Newmet Koch	rod
Li	Merck	chunk
Mg	Newmet Koch	rod
Cu	Aldrich	granular
Pd	MaTecK	foil
Ag	Newmet Koch	granular
In	Newmet Koch	chunk

* R = Sc, Y, La–Nd, Sm, Gd-Lu

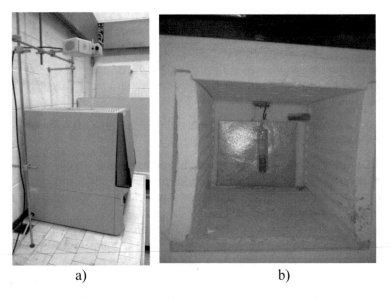

a)   b)

Fig. 2.1 a Resistance furnace equipped with rotation machine; **b** Ta crucibles (sealed in quartz phial) positioning inside the furnace

crystals of good quality and size, suitable for further structural studies, were obtained. This method was also useful to favour a better dissolution of the constituting elements inside the flux. Selective oxidation and/or centrifugation at $T > T_m(\text{flux})$ were used as separation methods.

2.2 Microstructure and Phase Analysis

In order to perform the metallographic analysis, samples were generally embedded in a phenolic resin with carbon filler, by using the automatic hot compression mounting press Opal 410 (ATM GmbH, Germany). Samples synthesized by flux method were analysed prior products separation: to avoid the metal solvent melting, they were embedded in a cold-curing resin, conductive due to the presence of copper filler.

The automatic grinding and polishing machine Saphir 520 (ATM GmbH, Germany) was used to obtain smooth alloy surfaces suitable for microscopic examinations. Grinding was performed by means of SiC papers with grain size decreasing from 600 to 1200 mesh, using running water as lubricant; for polishing, diamond paste with particle size decreasing from 6 to 1 μm was used with an alcohol based lubricant. Petroleum ether was employed to clean samples ultrasonically after each polishing step.

In the case of flux prepared samples only the 1200 mesh SiC paper was used applying a low pressure, due to In ductility.

2.2.1 Light Optical Microscope Analysis (LOM)

Samples surface observation after each polishing step, preliminary microstructure examination and single crystal selection for the further X-ray analysis were performed with a light microscope Leica DM4000 M (Leica Microsystems Wetzlar GmbH, Welzlar, Germany), achieving magnifications in the range 50–1000X.

Two different optical microscopy illumination techniques were employed: brightfield (BF) and darkfield (DF) mode. The former was applied for sample surfaces observation and microstructure examination; the latter for selecting single crystals suitable for the subsequent X-ray analysis. The substantial difference between the brightfield *and* darkfield modes is due to the illuminating system (Fig. 2.2).

In the BF mode the incident beam is perpendicular to the sample and all the light reflected by the latter is collected by the objective lens. On the obtained image, different areas of the sample surface appear brighter or darker depending on their degree of absorption of the incident light. This technique is then suitable for observing the sample microstructure, provided that there is sufficient contrast between phases.

In the DF mode the angle between incident beam and sample is close to 180°, therefore the major part of the incident beam is reflected away and only the fraction of light scattered by the sample is collected by the objective lens. As a result, in the DF image the specimen irregularities, able to scatter light, appear bright against a black background. This technique was then used to examine the mechanically fragmented alloys with the aim to select single crystals suitable for the structure determination (size, shape, borders and surfaces were particularly taken into account for this purpose).

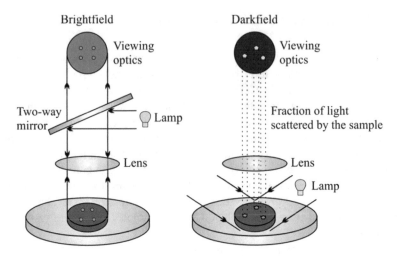

Fig. 2.2 Schematic representation of brightfield and darkfield optical microscopy illumination techniques

2.2.2 Scanning Electron Microscopy (SEM) and Energy Dispersive X-ray Spectroscopy (EDXS)

Microstructure examination as well as qualitative and semi-quantitative analyses were performed by a scanning electron microscope (SEM) Zeiss Evo 40 (Carl Zeiss SMT Ltd, Cambridge, England) equipped with a Pentafet Link Energy Dispersive X-ray Spectroscopy (EDXS) system managed by the INCA Energy software (Oxford Instruments, Analytical Ltd., Bucks, U.K). In the scanning electron microscope a focused beam of high-energy electrons generates a variety of signals at the surface of solid specimens. The incident electron beam, source of the so-called "primary electrons", is thermoionically generated by a tungsten filament cathode. These electrons are accelerated by a high voltage and then focused by magnetic lenses (Fig. 2.3a). When the primary electron beam interacts with sample atoms it produces various signals (Fig. 2.3b) containing information about the sample surface topography and composition. The main signals are:

Secondary electrons (SE): they are generated by inelastic scattering between the primary beam and the valence electrons of the specimen atoms. These low energy (<50 eV) electrons are ejected from their orbitals and detected. Only the SE generated at the top surface (1–10 nm depth) of the specimen are emitted outside. Thus, this signal can be used to obtain topological information.

Back-scattered electrons (BSE): they are high energy electrons back scattered out of the sample by elastic interactions of the primary beam with specimen atoms. Since heavy elements (high atomic number) backscatter electrons stronger than light elements, they appear brighter in the image. Therefore, BSE are used to construct images with contrast between areas with different chemical compositions (the so

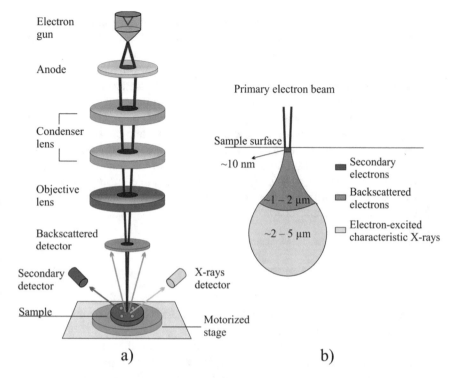

Figure 2.3 a Schematic representation of a SEM equipped with EDXS probe. **b** The interaction zone: the different signals and their interacting volume are shown

called "compositional contrast" or "Z-contrast"). The BSE images permit to identify the number of phases present in the analysed alloys.

X-ray: as a result of the inelastic scattering interactions core vacancies are generated. To fill these vacancies, electrons from higher energy levels fall down emitting X-rays. The released X-rays are characteristic for each element and can be used for qualitative and quantitative analysis by means of an EDXS probe, discriminating them on the basis of their energy. In the obtained emission spectra the peak positions give qualitative information; their intensity is proportional to elements concentration. To improve the accuracy of EDXS measurements, data are corrected by the so-called ZAF method (implemented in INCA software) where Z, A and F are, respectively, atomic number, absorption and fluorescence correction.

2.3 X-ray Diffraction (XRD)

X-rays are high energy electromagnetic radiation characterized by small wavelength ranging from about 0.1 to 100 Å. The oscillating electromagnetic field associated with X-rays interacts with the sample electrons forcing them to oscillate with the same frequency. The oscillating electrons emit radiation with the same frequency as the incident beam. Hence, X-rays are elastically scattered by electrons, obeying the Thompson law. Considering an isolated atom, the X-ray scattered from its electrons can give constructive or destructive interferences, depending on the scattering angle. The function that describes this behaviour is the atomic scattering factor (f_j for a generic atom j). In a crystal (*i.e.* an ordered, periodic arrangement of atoms), each atom acts as a source of scattered X-rays; these waves add constructively or destructively, depending on the direction of the diffracted beam and the atomic positions. The analysis of the diffracted radiation under convenient experimental conditions can therefore give information on the crystal structure of the examined material.

The structure factor (F_{hkl}) is a mathematical function describing the amplitude and phase of a wave diffracted from crystal lattice planes characterized by Miller indices h,k,l [1]. For a unit cell containing n atoms j, the structure factor for the reflection hkl is:

$$F_{hkl} = \sum_{j=0}^{n} f_j e^{2\pi i \left(hx_j + ky_j + lz_j\right)} \tag{2.1}$$

where f_j is the atomic scattering factor, *(hkl)* are the Miller indices and (x_j, y_j, z_j) are the coordinates of the atom j.

In Eq. (2.1) the atomic scattering factor f_j depends on the chemical nature of atoms, and the term $2\pi(hx_j + ky_j + lz_j)$ expresses the phase of the structure factor, depending on the positions of atoms within the unit cell. The common geometrical representation of the structure factor is shown in Fig. 2.4.

During an X-ray diffraction experiment the only information that can be measured is the intensity and position of the diffracted radiation. The structure factors *F(hkl)* are directly related to the integrated intensity I*(hkl)* of the corresponding reflection h,k,l ($I_{hkl} \propto |F_{hkl}|^2$). Therefore, from the measured intensities, we can calculate the magnitude of the structure factor but not its phase (Fig. 2.4). This drawback, known as the "phase problem", is the main problem of crystallography. During last century

Fig. 2.4 Representation of the structure factor F_{hkl} on the Argand-Gauss plane. Magnitude and phase angle $\Phi(hkl) = 2\pi(hx_j + ky_j + lz_j)$, are visible

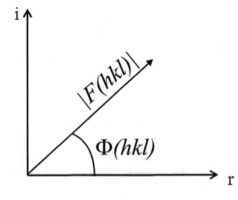

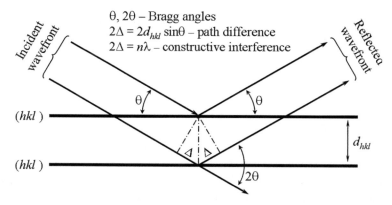

Fig. 2.5 Diffraction of X-rays from crystal lattice planes illustrating Bragg's law. Reproduced with permission from [2]

lots of methods (i.e. Patterson synthesis, direct method, charge-flipping, etc.) were developed to circumvent this problem.

The knowledge of the structure factor (both module and phase) is essential to solve a crystal structure, since it allows determining the electron density distribution function ($\rho(x, y, z)$), the maxima of which correspond to atomic positions. In fact, the representation that connects $\rho(x, y, z)$ to the diffraction pattern is:

$$\rho(x, y, z) = \frac{1}{V} \sum_{hkl} F_{hkl} e^{-2\pi i\left(hx_j + ky_j + lz_j\right)} \qquad (2.2)$$

Despite this complexity, the conditions required for diffraction can be understood by means of the simple Bragg's law (Eq. 2.3), considering diffraction as a reflection at the hands of crystal lattice plane (Fig. 2.5).

$$2d_{hkl} \sin \theta = n \cdot \lambda \qquad (2.3)$$

where:

d_{hkl}: distance between plane $\{hkl\}$	$2 \cdot d_{hkl} \cdot \sin\theta$: path difference
hkl: Miller indices	n: diffraction order
θ: diffraction angle	λ: wavelength of the X-rays

During this work two different X-ray techniques were employed:

1. X-ray powder diffraction for phase identification and calculation of lattice parameters and Rietveld structure refinement for selected phases.
2. X-ray single crystal diffraction to solve the crystal structure of new compounds.

For our purposes, the X-ray powder diffraction data can be conveniently treated on the basis of the Bragg's law. The task of solving new crystal structures is much more complicated, since it needs to solve the aforementioned "phase problem" yielding to crystal electron density.

Both the X-ray powder and single crystal diffractometer are constituted by the following principal parts:

(1) X-ray source
(2) Monochromator
(3) Sample holder
(4) Detector

The most common X-ray source is a cathodic vacuum tube (Fig. 2.6) containing a W cathode and a metallic anode (generally Cu for powder or Mo for single crystal experiments). Under high voltage conditions, the cathode thermoionically emits electrons, which collide against the anode material. Some of these electrons have sufficient energy to create electron vacancies in the anode core shell. Thus, electrons from higher energy levels fall down emitting characteristic X-rays.

The emitted X-rays should be further monochromatized with the aid of proper X-ray filters or monochromator crystals

The detector function is to record intensities and positions of the diffracted radiation.

Fig. 2.6 Schematic representation of an X-ray tube

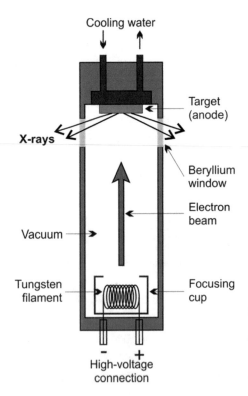

In modern detectors the signal, which is usually an electric current, is easily digitized and transferred to a computer for further processing and analysis. In general, detectors could be divided into three categories on the basis of their capability to detect the direction of the beam in addition to counting the number of photons: point, line and area detectors. The first one registers only the intensity of the diffracted beam, one point at a time, *e.g.* scintillation detector, solid-state detector. The second supports spatial resolution in one direction while the last one supports resolution in two dimensions, *e.g.* Positive Sensitive Detectors (PSD) and Charge Coupled Devices (CCD).

The X-ray powder diffractometer used during this work was equipped with a scintillation detector, while X-ray single crystal diffractometer with a CCD detector.

A typical scintillation detector is constructed from a crystal scintillator coupled with a photomultiplier tube. CCD detectors are photon counters, solid state devices that accumulate charge (electrons) in direct proportion to the amount of light that strikes them, allowing collecting many reflections at once. A typical image of CCD data is shown in Fig. 2.7a). The Rocking curve of the selected peak could be plotted separately (see the red zone in Fig. 2.7b) in this way the profiles of peaks could be examined. What is more, a 3D image of wanted zone can be visualized (see green area in Fig. 2.7c).

2.3.1 X-ray Powder Diffraction

X-ray diffraction on powder samples was performed by means of a diffractometer Philips *X'Pert* MPD (Cu *Kα* radiation, $\lambda = 1.5406$ Å, graphite crystal monochromator, scintillation detector, step mode of scanning), with a $\theta{:}2\theta$: Bragg–Brentano geometry. In this geometry (see Fig. 2.8) the X-ray source is stationary while the sample and the detector rotations are synchronized to fulfil the $\theta{:}2\theta$ requirement. X-ray lamp, sample holder and detector are located on the "focusing-circle". In order to avoid X-ray angular divergence, which would give broad and asymmetric peaks, the radiation is collimated by means of two slits (S1 and S2) placed after the X-ray lamp and before the monochromator.

Sample powders of suitable dimensions (between about 10 and 50 μm) and statistically oriented in all directions were obtained by mechanical grinding using an agate mortar and pestle. Powders were pressed in the cavity of a monocrystalline silicon flat sample designed in order to minimize background and zero-shift effects.

Measured powder patterns were collected in $10°–100°\ 2\theta$ range, with a scanning step of *ca.* $0.02°$ with a time per step varying from 10 to 20 s.

The experimental powder diffraction pattern could be interpreted as a set of discrete peaks, called Bragg reflections, superimposed over a continuous background [2]. Disregarding the background, the peaks may be described by their positions, intensities and shapes which contain information about the crystal structure, specimen property and instrumental parameters, as shown in Table 2.2.

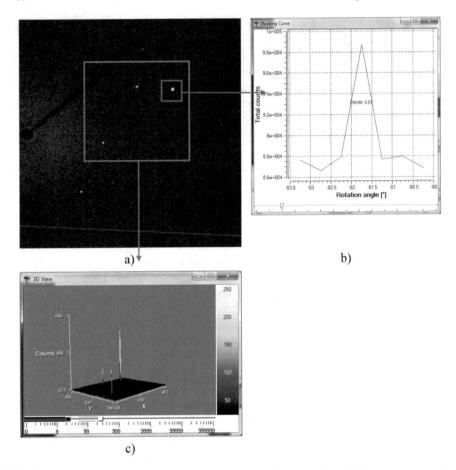

Fig. 2.7 A typical CCD image (**a**) accompanied with a rocking curve (**b**) and 3D view of a selected zone (**c**)

As it is clear from Table 2.2, in order to deduce wanted crystallographic parameters, the access to peak positions and intensities is essential.

When dealing, as in this work, with multi-phase sample the measured patterns is the sum of the characteristic diffraction patterns of the constituent crystalline phases.

For all samples the phase identification was performed. It consists of comparing the positions and relative intensities of the experimental peaks with the calculated diffraction patterns of phases expected in the sample. A theoretical diffraction pattern can be calculated only if structural data are known from the literature or from a structural model obtained after single crystal X-ray diffraction analysis. During this work, phase identification was performed with the help of the software PowderCell [3].

After that, the correct (*h,k,l*) Miller indices were assigned to peaks of each phase (indexing procedure) just on the basis of their positions. Further, lattice parameters

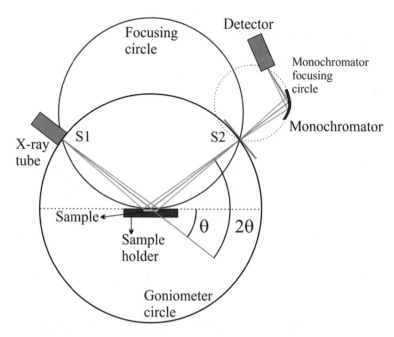

Fig. 2.8 Scheme of a powder diffractometer with a θ:2θ Bragg–Brentano geometry equipped with a curved monochromator

Table 2.2 Powder diffraction pattern components as a function of some crystal structure, specimen and instrumental parameters

Pattern component	Crystal structure	Specimen property	Instrumental parameters
Peak position	**Unit cell parameters** $(a, b, c, \alpha, \beta, \gamma)$	Absorption, porosity	Radiation (λ), instrument/sample alignment
Peak intensity	**Atomic parameters** $(x, y, z, B,$ etc....$)$	Preferred orientation, absorption	Radiation (Lorentz, polarization)
Peak shape	Disorder, defects, crystallinity	Grain size, strain, stress	Radiation (spectral purity), geometry, Beam conditioning

Structural parameters are in bold. Adapted with permission from [2]

were refined by a least-squares method implemented in the software LATCON [4]. In fact, lattice parameters are related to h,k,l triplets and the inverse square of the interplanar distance $1/d^2$ (obtained from Bragg's law) through the general formula:

$$\frac{1}{d^2} = \left[\begin{array}{l} \dfrac{h^2}{a^2 \sin^2 \alpha} + \dfrac{2kl}{bc}(\cos \beta \cos \gamma - \cos \alpha) + \dfrac{k^2}{b^2 \sin^2 \beta} \\[2mm] \quad + \dfrac{2hl}{ac}(\cos \alpha \cos \gamma - \cos \beta) + \dfrac{l^2}{c^2 \sin^2 \gamma} + \dfrac{2hk}{ab}(\cos \alpha \cos \beta \\[2mm] \quad - \cos \gamma) \\[2mm] /\left(1 - \cos^2 \alpha - \cos^2 \beta - \cos^2 \gamma + 2\cos \alpha \cos \beta \cos \gamma \right) \end{array} \right]$$

$$(2.4)$$

where a, b, c are the lattice parameters and α, β, γ are the angles between them. This general formula refers to the triclinic crystal system where a total of six independent parameters are required to define the unit cell dimensions. A similar relation could be easily obtained for other crystal systems, reducing the number of variables. For example, for the cubic crystal system (i.e. a = b = c, $\alpha = \beta = \gamma = 90°$) the formula is:

$$\frac{1}{d^2} = \frac{\left(h^2 + k^2 + l^2\right)}{a^2} \tag{2.5}$$

In this work, a set of measured peak positions was used to precisely calculate lattice parameters by a least square regression procedure with the help of the *LATCON* software [4].

For some almost single phase samples a crystal structure refinement, based on the procedure developed by Rietveld [5], was performed by means of FullProf software [6]. At the end of this procedure not only the unit cell parameters, but also the atomic ones (both unit cell content and spatial distributions of atoms within the cell) are obtained. In order to successfully apply this method, the access to high quality X-ray powder patterns (high intensity peaks and low background), depending on both the sample and the experimental set up, it's mandatory. In addition, it is also essential to have an adequate structural model as a starting point to generate a calculated profile which, when fully refined, should closely resemble the collected one. Mathematically, this is done through a non-linear least squares algorithm applied to the following function:

$$S = \sum_{i}^{n} w_i \left(Y_i^{obs} - Y_i^{calc}\right)^2 \tag{2.6}$$

where Y_i^{obs} is the observed and Y_i^{calc} is the calculated intensity of a point i of the powder diffraction pattern and w_i is the weight assigned to the ith data point. The summation is carried over all the n measured data points. The Y_i^{calc} is expressed as a function of the background and the parameters related to the pattern component (see Table 2.2) which have to be successfully refined one after the other. In particular, the refinement procedure followed in this work was the following: scale factor, background fitted by linear interpolation of a set of points (~60) taken from the collected spectrum, sample displacement and zero shift, unit cell dimensions, peak shape function parameters like full width at half maximum and asymmetry, atomic coordinates of all independent atoms, site occupation and atomic displacement parameters. To represent the peak shape a pseudo-Voigt curve was selected and refined.

Similar to single crystal data (see 2.3.2) the quality of the performed refinement is quantified by the following figures of merit:

the profile residual R_p,

$$R_p = \frac{\sum_{i=1}^{n} \left| Y_i^{obs} - Y_i^{calc} \right|}{\sum_{i=1}^{n} Y_i^{obs}} \cdot 100 \tag{2.7}$$

the weighted profile residual R_{wp},

$$R_{wp} = \left[\frac{\sum_{i=1}^{n} w_i \left(Y_i^{obs} - Y_i^{calc} \right)^2}{\sum_{i=1}^{n} w_i (Y_i^{obs})^2} \right]^{\frac{1}{2}} \cdot 100 \tag{2.8}$$

the Bragg residual R_B,

$$R_B = \frac{\sum_{i=1}^{m} \left| I_j^{obs} - I_j^{calc} \right|}{\sum_{i=1}^{m} I_j^{obs}} \cdot 100 \tag{2.9}$$

the expected profile residual R_{exp},

$$R_{exp} = \left[\frac{n - p}{\sum_{i=1}^{n} w_i (Y_i^{obs})^2} \right]^{\frac{1}{2}} \cdot 100 \tag{2.10}$$

and the goodness of fit χ^2

$$\chi^2 = \frac{R_{wp}}{R_{exp}} = \frac{\sum_{i=1}^{n} w_i \left(Y_i^{obs} - Y_i^{calc} \right)^2}{n - p} \cdot 100 \tag{2.11}$$

With n is the total number of points measured, p is the number of free least square parameters, m is the number of independent Bragg reflections and I_j^{obs} is the observed and I_j^{calc} the calculated integrated intensity of the jth Bragg peak. Generally, a good fit is indicated by low values of the residuals. To a certain degree, the good quality of the results is mainly established through χ^2 which is, in the ideal case, equal to unity when the obtained and the expected values are the same within the estimated error on each measured datum. Anyway, it is important to underline that none of the residuals could be a substitute for the plots of the observed and calculated powder diffraction patterns, supplemented by the difference $\Delta Y_i = Y_i^{obs} - Y_i^{calc}$, represented on the same scale. This is very important since it could happen that good values of the residuals are not associated with good refinement, as revealed by the plot. For example, in the extreme case that the background assumes particularly high values, since R_p and R_{wp} absorb its contribution, the corresponding figure of merit would be excellent having large values at the denominator. That is why also the aforementioned plots must be always reported. In this work, Rietveld refinement was performed by means of FullProf software [7].

2.3.2 X-Ray Single Crystal Diffraction

A full-sphere dataset was obtained in routine fashion at ambient conditions on a four-circle Bruker Kappa APEXII CCD area-detector diffractometer equipped by the graphite monochromatized Mo $K\alpha(\lambda = 0.07137\text{Å})$ radiation, operating in ω-scan mode. Crystals, glued on glass fibres, were mounted in a goniometric head and then in the goniostat inside diffractometer camera (Fig. 2.9). This device allows automated and highly precise movement of the crystal into almost any orientation with respect to X-ray incident beam and the detector. Crystal orientation is specified by a set of angles in which individual reflections are directed to a CCD area detector for intensity measurement. This set of angles allows rotation of the goniometric head (angle ϕ) movement of the head around a circle centred on the X-ray beam (angle χ), and rotation of the χ circle around an axis perpendicular to the beam (angle ω). The fourth angle define the detector position with respect to the beam (angle θ).

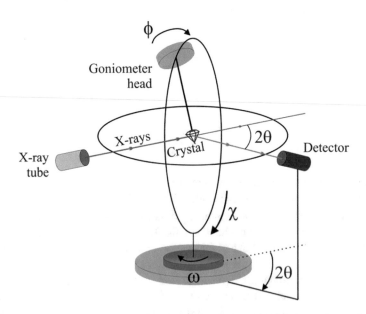

Fig. 2.9 Simplified scheme of a four-circle diffractometer. Labels highlight rotation angle system and instrumental devices

Thus, the measurement result is a set of reflections with different relative intensities. Before data collection it is convenient to collect some frames and harvest reflections for a preliminary indexing. In this way the Bravais type of lattice and crystal class of symmetry can be evaluated. Moreover, at this point one can discard bad quality crystals. After that, the full measurement strategy can be performed. The acquired scans (time of exposure of 20–30 s per frame collecting intensity over reciprocal space up to 30° in θ) were integrated using SAINT [8] and the highly redundant final dataset was corrected for Lorentz and polarization effects. Empirical absorption corrections are strictly necessary to evaluate the contribution of the structure factor $|F^2|$ to the measured intensity as precisely as possible. In the present work, to take them in considerations, the SADABS software [9] (TWINABS for twinned crystals) was applied to all data further merged to acceptable first (R_{int}) and second (R_{sigma}) merging residual values:

$$R_{int} = \frac{\sum \left| F_o^2 - \langle F_o^2 \rangle \right|}{\sum F_o^2} \tag{2.12}$$

$$R_{sigma} = \frac{\sum \sigma F_o^2}{\sum F_o^2} \tag{2.13}$$

where F_o is the measured module of the structure factor and σ is the estimated standard uncertainty. The R_{int} and R_{sigma} values allow evaluating the quality of the recorded dataset prior to structure solution; acceptable values are normally less than 0.1.

The crystal structure models were obtained in a few iteration cycles by applying the charge-flipping algorithm implemented in *JANA2006* [10] and then refined by full-matrix least-squares procedures on $|F^2|$ using SHELX-97 software package [11]. Other softwares were used during structure solution and its refinement [12].

To judge how well the obtained model fits the observed data the residual factors are considered. The three most commonly used *R*-factors are:

the unweighted residual factor *R* (*R*1 in SHELX),

$$R1 = \frac{\sum ||F_o| - |F_c||}{\sum |F_o|} \tag{2.14}$$

the weighted residual *wR* (*wR*2 in SHELX),

$$wR2 = \left[\frac{\sum w \left(F_o^2 - F_c^2 \right)^2}{\sum w \left(F_o^2 \right)^2} \right]^{\frac{1}{2}} \tag{2.15}$$

and the goodness of fit GoF,

$$GoF = \left[\frac{\sum w \left(F_o^2 - F_c^2 \right)^2}{(N_R - N_P)} \right]^{\frac{1}{2}} \qquad (2.16)$$

where $|F_o|$ and $|F_c|$ are respectively the observed and calculated module of the structure factor, w is the weighting factor individually derived from the standard uncertainties of the measured reflections, N_R is the number of independent reflections and N_P is the number of refined parameters. Normally, the $wR2$ value is twice as big as the $R1$; the absolute value of both should not exceed 0.15. The goodness of fit (GoF) should approach the unity (ideal value). At the end, even if residuals values are excellent, the difference Fourier map was accurately checked. Being the difference between the observed and calculated electron densities ($\rho_o - \rho_c$), obtained respectively from observed and calculated structure factors, it should be as flat as possible indicating that the obtained structure is a good model of the real one.

Experimental details and crystallographic data for studied germanides are reported in the following.

2.4   Differential Thermal Analysis (DTA)

Differential thermal analysis (DTA) was applied to a few samples, in order to gain insights on the formation of selected intermetallics. Measurements were carried out using a LABSYS EVO (SETARAM Instrumentation, Caluire, France) equipped with type S (Pt-PtRh 10%) thermocouples, in the temperature range 25–1100 °C using custom-made tantalum crucibles. The sample crucibles were arc-sealed under inert atmosphere after having cooled it in liquid nitrogen, so as to avoid undesired reactions. Crucibles of the same weight as the sample containers were used as references. The DTA curves were recorded under a continuous flow of argon (20 mL/min) to avoid oxidation of the crucibles at high temperatures. Different thermal cycles were applied depending on the sample. The obtained thermograms were evaluated with the software Calisto, supplied by SETARAM with the DTA equipment. Prior and after the DTA experiments the samples were characterized by SEM/EDXS and/or XRPD.

References

1. Internet website of IUCr: https://reference.iucr.org/dictionary/Structure_factor (15/11/2018)
2. Pecharsky V, Zavalij PY (2003) Fundamentals of powder diffraction and structural characterization of materials. Springet Science+Business Media, LLC, New York
3. Kraus W, Nolze G (1996) J Appl Crystallogr 29:301–303
4. Schwarzenbach D (1996) LATCON: refine lattice parameters. University of Lausanne, Lausanne, Switzerland
5. Rietveld HM (1969) J Appl Cryst 2:65

6. Rodriguez-Carvajal J (1993) J Phys B 192:55–69
7. Rodriguez-Carvajal J (2001) Recent developments in the program FullProf. Newsletter 26:12–19
8. Bruker. AXS Inc., SAINT, Bruker. AXS Inc., Madison, WI, USA, **2011**
9. Bruker. AXS Inc., SADABS, Bruker. AXS Inc., Madison, WI, USA, 2011
10. Petricek V, Dusek M Palatinus L (2014) Z. Kristallogr. 229(5):345–352
11. Sheldrick GM (1997) SHELXS-97, Program for crystal structure refinement". University of Gottingen, Germany
12. Bruker, APEX2, SAINT-Plus, XPREP, SADABS, CELL_NOW and TWINAB*S*. Bruker AXS Inc., Madison, Wisconsin, USA, **2014**.

Chapter 3
Theoretical Methods

During the twentieth century an impressive progress in the theories and computational methods was achieved so that nowadays it is possible to obtain/predict energies and properties of many solid systems with high accuracy, assuming a model of the solid as a starting point. Anyway, the development of a theoretically based predictive approach which could guide the preparation of new compounds with selected properties is still one of the main dream of both solid state chemists and physicists. It is worth to note that some strategies were already developed. The most widespread is based on the calculated energy differences between various structural models, generated on the basis of literature data or some specific algorithms [1]. This method was also used during this work for the family of the R_2MGe_6 phases (see Chap. 4). Nevertheless, mainly from the chemical perspective, the reason why the energy of one compound with a certain structure is lower/higher than another is an important explanation that should be accounted for also invoking chemical concepts. In this framework it is evident that a connection between the formal quantum mechanics language based on the band structure theory, which operates in the reciprocal space, and chemical concepts have to be found. One powerful approach to obtain the bonding information from the results of the electronic band structure calculations (regardless of wave-function or density-based) is the Crystal Orbital Hamilton Population (COHP) method [2, 3], which was applied to some synthesized intermetallic (see Chap. 6). In the COHP technique, bonding, non-bonding, and antibonding interactions are identified for pairs of atoms in a given solid-state material. Despite its power, the COHP method does not permit to interpret results in terms of classical chemistry concepts, such as atomic volumes, effective charges, oxidation numbers, effective covalent bond orders, bonds and lone pairs which is allowed by the position-space chemical bonding techniques. The latter were recently applied, for the first time, to predict new Half-Heusler phases whose existence was subsequently confirmed by experiments [4].

© The Editor(s) (if applicable) and The Author(s), under exclusive license
to Springer Nature Switzerland AG 2020
R. Freccero, *Study of New Ternary Rare-Earth Intermetallic Germanides
with Polar Covalent Bonding*, Springer Theses,
https://doi.org/10.1007/978-3-030-58992-9_3

In this work the aforementioned methods were applied, using different codes, with the main purpose to elucidate chemical bonding in the synthesized germanium-rich intermetallic compounds.

3.1 Density Functional Theory (DFT)

Advances in the power and speed of computers have made density functional theory (DFT) [5] calculations an almost routine procedure on a PC [6]. From these calculations the equilibrium geometry, the energy, and the wave function of a molecule or a crystal can be determined. From the wave function one can obtain all the properties of the compound, including the distribution of electronic charge, or electron density, which gives very useful information on bonding and geometry.

3.1.1 The Schrödinger Equation

In quantum mechanics, all the information about a system is included in its wave function (Ψ). Without taking into account relativistic and time dependency effects, the wave function can be obtained by solving the Schrödinger equation [7]:

$$\hat{H}\,\Psi = E\,\Psi \tag{3.1}$$

where E is the system total energy and $\hat{H}$ is the Hamiltonian operator. Schrödinger equation states that the total energy can be found applying the Hamiltonian operator to the wave function. The Hamiltonian operator contains different contributions: kinetic energy for electrons and nuclei, attractive nucleus-electron electrostatic interactions, repulsive electron-electron and nucleus-nucleus electrostatic interactions. The system Hamiltonian can be simplified by the Born-Oppenheimer approximation. According to it, since electrons kinetic energy is higher than that for nuclei (due to their large mass difference) one can consider nuclei as static positive charges with zero kinetic energy. As a consequence, the term which accounts for the nucleus-nucleus repulsion can be considered to be a constant. According to the mathematical rules, the addition of a constant to an operator simply adds to the eigenvalues (E in Schrödinger equation) but leaves unaffected the eigenfuntions (Ψ). So, this term can be neglected. For a system composed by N electrons and M nuclei, the electronic Hamiltonian ($\hat{H}_{elec,}$) can be expressed as:

$$\hat{H}_{elec} = -\frac{1}{2}\sum_{i=1}^{N}\nabla_i^2 - \sum_{i=1}^{N}\sum_{A=1}^{M}\frac{Z_A}{r_{iA}} + \sum_{i=1}^{N}\sum_{j>1}^{N}\frac{1}{r_{ij}} = \hat{T} + \hat{V}_{Ne} + \hat{V}_{ee} \tag{3.2}$$

where $\hat{T}$ is the kinetic energy operator, $\hat{V}_{Ne}$, is the nucleus-electron attractive energy operator and $\hat{V}_{ee}$ is the electron-electron repulsive energy operator. The solution of Schrödinger equation for $\hat{H}_{elec}$ is given by the electronic wave function and the electronic energy. Unfortunately, even simplifying the equation, an exact solution can be found only for systems with one nucleus and one electron, the so-called hydrogen-like systems (e.g. H, He^+, Li^{2+}, etc.). In other cases, it is possible only to find approximate solutions to the Schrödinger equation. During the twentieth century a lot of accurate approximations were proposed. Among them the most common are the one-electron approximation, like Hartree-Fock, in which the many-body wavefunction Ψ is a product of antisymmetrized one-electrons functions (orbitals). The easiest way to generate such a function is through a Slater determinant.

3.1.2 Electron Density

Since the N-electrons wave function depends on $4N$ variables (3 spatial and 1 spin coordinates for each electron), it is complicate to find accurate approximate solutions for large systems, as solids and biomolecules are. Moreover, increasing system complexity, data interpretation becomes harder and harder. If considering the electron density, $\rho(r)$, which is a real space quantity depending only on three coordinates, the situation becomes somewhat simpler.

The electron density (in spherical coordinate system) is described by the following formula:

$$\rho(r) = N \int d\sigma_1 \int dx_2 \ldots \int dx_N |\Psi(r_1\sigma_1, \, x_2, \ldots x_N)|^2 \qquad (3.3)$$

r = electron spatial coordinate
σ = electron spin coordinate
$x = (r, \sigma)$

It is defined as the integral of the wave function over the spin coordinates of all electrons and over all but one of the spatial variables multiplied by the number of electrons (N).

As a consequence, $\rho(r)$ determines the N-probability to find *any* of the N electrons in the volume dx_1 whereas the remaining N-1 electrons are anywhere in space, regardless of spin. Thus, the maximum value this quantity can attain is N, which is why it is not probability in the strict mathematical sense (*i.e.* a true probability function must be normalized to unity). The electron density is a non-negative function of only the three spatial variables that has a maximum at the position of each nucleus and decays rapidly away from these positions, vanishing at infinity and integrates to the total number of electrons:

$$\rho(r \to \infty) = 0 \quad \int \rho(r)dr = N \qquad (3.4)$$

The same approach can be applied to define the electron pair density, $\rho_2(r_1, r_2)$:

$$\rho_2(r_1, r_2) = N(N-1) \int d\sigma_1 \int d\sigma_2 \int dx_3 \dots \int dx_N |\Psi(r_1\sigma_1, r_2\sigma_2, x_3, \dots x_N)|^2$$

(3.5)

This function describes the "total probability" to find a pair of electrons with spins σ_1 and σ_2 simultaneously within two volume elements dr_1 and dr_2, while the remaining N–2 electrons have arbitrary positions and spins. It is of great importance since it contains all information about electron correlation.

3.1.3 Hohenberg-Kohn Theorems

As said above, by means of electron density it is possible to overcome the problem of finding the complicated wave function, which depends on $4N$ variables, especially for extended systems, like solids. As shown in Eq. (3.2), we can write the electronic Hamiltonian operator as the sum of three different contributions:

$$\hat{H}_{elec} = \hat{T} + \hat{V}_{Ne} + \hat{V}_{ee}$$

The term $\hat{V}_{Ne}$, also called external potential, is system dependent, while $\hat{T}$ and $\hat{V}_{ee}$ are the same for any N electrons system. So, we can uniquely determine electron density from the external potential. The question is: is it possible to uniquely determine the external potential from electron density?

With the purpose to solve Schrödinger equation using $\rho(r)$, Hohemberg and Khon in 1964 [5], provided two theorems, which are the basis of *DFT*.

The first Hohemberg-Kohn theorem, quoting literally [5], states that *the external potential is (to within a constant) a unique functional of $\rho(r)$; since in turn external potential fixes the Hamiltonian we see that the full many body particle ground state is a unique functional of $\rho(r)$"*.

Thus, we can write total ground state energy as a ground state density functional:

$$E[\rho] = V_{Ne}[\rho] + T[\rho] + V_{ee}[\rho] = \int \rho(r) V_{Ne}(r) dr + F_{HK}[\rho]$$

(3.6)

where $F_{HK}[\rho]$ is the Hohenberg-Kohn functional and contains the functional for the kinetic energy, $T[\rho]$, and that for the electron-electron interaction, $V_{ee}[\rho]$. If $F_{HK}[\rho]$ were known the Schrödinger equation could be solved exactly. Unfortunately, we are not able to find it. However a classical Coulomb part, $J[\rho]$, can be extracted from:

$$V_{ee}[\rho] = \frac{1}{2} \int \int \frac{\rho(r)\rho(r')}{|r-r'|} dr dr' + E_{ncl} = J[\rho] + E_{ncl}[\rho]$$

(3.7)

where $E_{ncl}[\rho]$ is the non-classical contribution to the electron-electron interaction.

Now the question is: how can we be sure that a guessed density is the ground state density we are looking for?

The second Hohemberg-Kohn theorem answers to latter question stating that $F_{HK}[\rho]$, *the functional that delivers the ground state energy of the system, delivers the lowest energy if and only if the input density is the ground state density.* This theorem is a variational principle, so, for any trial density the energy obtained is an upper limit to the true ground state energy:

$$E_0 \leqslant E[\rho] = T[\rho] + V_{Ne}[\rho] + J[\rho] + E_{ncl}[\rho] \tag{3.8}$$

It's important to highlight that the applicability of this variational principle is limited to the ground state. Hence, we cannot easily transfer this strategy to the problem of excited states.

3.1.4 Kohn-Sham Method

As stated above, the main problem is to find the expression for $T[\rho]$ and $E_{ncl}[\rho]$. In 1965, Kohn and Sham [8] suggested to calculate the exact kinetic energy of a non-interacting reference system, with the same density as the real one. The remaining contribution to the kinetic energy, a small part of the total, is treated by non-classical approximate functional. From Eqs. (3.6) and (3.7), it derives that:

$$F_{HK}[\rho] = T[\rho] + J[\rho] + E_{ncl}[\rho] \tag{3.9}$$

$J[\rho]$ is a classical term and can be exactly determined, but the other two terms are unknown. If now we consider a non-interacting system, we can modify Eq. (3.9):

$$F_{HK}[\rho] = T_s[\rho] + J[\rho] + E_{ncl}[\rho] + (T[\rho] - T_s[\rho]) = T_s[\rho] + J[\rho] + E_{xc}[\rho] \tag{3.10}$$

Where:

$T_s[\rho]$ is the kinetic energy functional of the non-interacting system
$E_{xc}[\rho]$ is the exchange and correlation energy which contains all the unknown terms.

$E_{xc}[\rho]$ still remains unknown and represents the main challenge of DFT. To calculate the exact kinetic energy for the non-interacting system Kohn and Sham proposed a Hartree-Fock like method (one-electron approximation, Slater determinant for the wave function) using the so called Kohn-Sham orbitals for the non-interacting system, (ψ_i), and a Hamiltonian operator, H_S, composed by kinetic energy and the so-called effective potential, $V_{eff}(r)$. This potential includes the external potential, the exchange and correlation part and also the classical term. The latter can't be J

since it is a two electrons quantity. Then, the effect of the electron-electron repulsion on a certain electron at position r should approximately be given by the electrostatic potential generated by all other electrons, on average, at position r (mean filed approximation). This is the Hartree potential:

$$V_H(r) = \int \frac{\rho(r')}{|r - r'|}\, dr' \qquad (3.11)$$

In this way $V_{eff}(r)$ can be defined only on basis of $\rho(r)$, and then the single particle eigenvalues problem for the aforementioned Hamiltonian H_S can be solved:

$$\left(-\frac{1}{2}\nabla^2 + V_{eff}(r)\right)\psi_i = \epsilon_i\psi_i \qquad (3.12)$$

Electron density is recovered from obtained Kohn-Sham orbitals together with the kinetic energy Ts:

$$T_S = -\frac{1}{2}\sum_i^N \langle \psi_i|\nabla^2|\psi_i\rangle \quad \rho_S(r) = \sum_i^N\sum_S |\psi_i(r, s)|^2 = \rho(r) \qquad (3.13)$$

It is important to highlight that Kohn-Sham orbitals have not a physical sense: they're used to obtain the exact electron density $\rho(r)$, which is equal to the electron density for the non-interacting system $\rho_S(r)$. In last decades many efforts were done to find approximate equations for the exchange and correlation potential, *e.g.* Local Density Approximation (LDA), Generalized Gradient Approximation (GGA) and hybrid functional (like B3LYP).

From the computational point of view, the Kohn-Sham equations (Eq. 3.11) are solved on the basis of the self-consistent field (SCF) method. It is necessary to construct the external potential $V_{Ne}(r)$ and start with a guessed electron density. Then the Hartree and exchange-correlation terms are computed allowing to have all the ingredients to build the effective potential V_{eff} and then H_S. Solving the Schrödinger Eq. (3.12) for all the electrons the Kohn-Sham orbitals are found leading to a new density according to (3.13). This process is repeated until convergence which means until the difference between the electron densities obtained at step n and n-1 is lower than a chosen value.It is also important to highlight that during this work almost all DFT calculation were performed on crystalline solids the periodicity of which is expressed by the lattice. Then, the Kohn-Sham equations are constructed in such a way that the Bloch's theorem is fulfilled [9]:

$$\psi(k,\ r + T) = e^{ikT}\psi(k, r) \qquad (3.14)$$

where T is one of the lattice vectors, $\psi(k, r)$ the crystal orbitals at a specific site r and k is the wave vector.

3.2 Quantum Theory of Atoms in Molecules (QTAIM)

Quantum Theory of Atoms In Molecules (QTAIM) was developed by Bader and co-workers in 1989 [10]. It is based on the electron density $\rho(r)$ obtained either from quantum mechanical calculations (see Sect. 3.1) or estimated experimentally from accurate X-ray diffraction data (see Sect. 2.3). Nowadays these kinds of experiment are more realistic, due to major technical advances, with the availability of fast area detectors, intense short wavelength synchrotron radiation and stable helium cooling devices.

The QTAIM theory is able to confer physical meaning to basic chemical concepts as atoms or bonds, via the topology of the gradient vector field of the electron density. The gradient vector field can be calculated using the gradient of electron density ($\nabla\rho(r)$) via the concept of gradient path, that is a curve to which the gradient vector is tangential at each of its points. Gradient paths have a beginning and an end, they never cross and they can meet only where $\nabla\rho(r) = 0$. Such a point is called a Critical Point (CP).

The critical points of $\rho(r)$, representing special positions, can be local minima, maxima or saddle points. They can be differentiated by means of the associated hessian matrix $H(r)$ which is a real, symmetric 3×3 matrix, formed by the pure and mixed second derivatives of $\rho(r)$:

$$H(r) = \begin{pmatrix} \frac{\partial^2\rho}{\partial x^2} & \frac{\partial^2\rho}{\partial x\partial y} & \frac{\partial^2\rho}{\partial x\partial z} \\ \frac{\partial^2\rho}{\partial y\partial x} & \frac{\partial^2\rho}{\partial y^2} & \frac{\partial^2\rho}{\partial y\partial z} \\ \frac{\partial^2\rho}{\partial z\partial x} & \frac{\partial^2\rho}{\partial z\partial y} & \frac{\partial^2\rho}{\partial z^2} \end{pmatrix} \qquad (3.15)$$

Unlike $\rho(r)$, the values of this matrix depend on the choice of the coordinates (x, y, z). To eliminate this arbitrariness, the mathematical procedure called matrix diagonalization is applied: the obtained eigenvalues (diagonal elements) give information on the curvature of $\rho(r)$. In QTAIM, a CP is denoted by (r, s) where the rank (r) is the number of non-zero eigenvalues and the signature (s) is the sum of the signs of the aforementioned eigenvalues. Types of CPs are listed in Table 3.1:Maxima in electron density $(3, -3)$ are also known as "attractors". In fact, due to its properties, the gradient path can be attracted to only this type of CPs. Critical point maxima can be:

Table 3.1 Types of CPs according to QTAIM analysis

Name	Acronym	(r, s)	CP type
(Non) nuclear attractor	(N)NA	$(3, -3)$	Maximum
Cage critical point	CCP	$(3, 3)$	Minimum
Ring critical point	RCP	$(3, 1)$	Saddle
Bond critical point	BCP	$(3, -1)$	Saddle

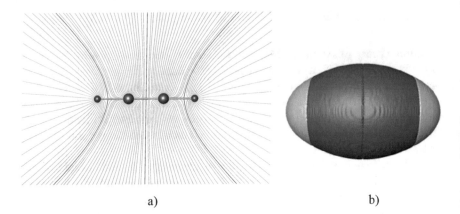

a)   b)

Fig. 3.1 a Representation of gradient vector fields (sets of infinite gradient paths) for the C_2H_2 molecule. Zero flux surfaces are shown in red; **b** Three-dimensional representation of atomic basins for the C_2H_2 molecule (the electron density is cropped at 0.02 e/bohr^3)

- Nuclear Attractors (NA), practically coinciding with the position of nuclei (black and blue spheres in Fig. 3.1a).
- Non nuclear attractors (NNA).

Minima in the electron density (3, 3) are Cage Critical Point (CCP) since they only occur in the centre of an atomic cluster. A saddle point can be of $(3, -1)$ or $(3, 1)$ types. The former is a Bond Critical Point (BCP) while the latter is a Ring Critical Point (RCP) since it is located at the centre of a cyclic molecule. The number of critical points must fulfil the Poincaré-Hopf theorem [11] which, for systems with periodic boundary conditions is:

$$\mathcal{N}(3, -3) - \mathcal{N}(3, -1) + \mathcal{N}(3, 1) - \mathcal{N}(3, 3) = 0 \qquad (3.16)$$

An infinite collection of gradient paths forms the gradient vector field (Fig. 3.1a). The gradient vector field is the basis for the partitioning of a molecular system into non-overlapping space filling regions, defined "atomic basins" (Fig. 3.1b). These basins are traversed by paths terminating at an attractor. The sum of a NA and its atomic basin is the QTAIM atom (the atomic basin for an atom X is labelled Ω_X).The boundary surfaces, called Interatomic Surfaces (IAS) between the QTAIM atoms must fulfil the zero-flux condition:

$$\nabla \rho(r) \cdot n = 0, \quad \forall r \in IAS \qquad (3.17)$$

The difference between the IAS and an arbitrary surface boundary is shown in Fig. 3.2 for the NaCl molecule: the normal to the IAS (1) is also perpendicular to $\nabla \rho$, which is not true for the arbitrary surface (2).Thanks to the zero-flux condition, it is possible to demonstrate that the QTAIM atoms can also be justified in a rigorous quantum

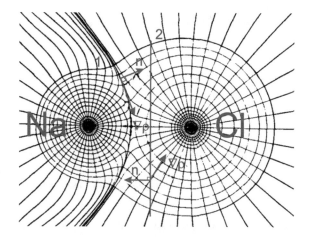

Fig. 3.2 Planar representation of gradient vector field and electron density contour map in the sodium chloride molecule focusing on the difference between the IAS (1) and an arbitrary surface (2). Adapted, with the permission of the Licensor through PLSclear, from [10]

mechanical sense as quantum subsystems with a well-defined electronic energy. Therefore, the total energy of a compound is the sum of atomic energies.

The integral of the electron density within an atomic basin yields the effective charge $Q^{eff}(\Omega_X)$ of atom X, through the following formula:

$$Q^{eff}(\Omega_X) = Z_X - \bar{N}(\Omega_X) = Z_X - \int_{\Omega_X} \rho(r)dr \qquad (3.18)$$

Where Z_X denotes the nuclear charge of atom X and $\bar{N}(x)$ is the average electronic basin population. An atom carrying a positive or a negative QTAIM effective charge may be called a QTAIM cation or anion.

3.3 Electron Localizability Indicator (ELI)

The Electron Localizability Indicator (ELI) was developed by M. Kohout in 2004 [12–15] and include a family of real space chemical bonding indicators based on the restricted population approach. Within this approach, two different position-dependent properties, called *control* and *sampling* property, are interconnected. The control property forms the basis for partitioning of space into compact, mutually exclusive and space-filling micro-cells with variable volumes. The sampling property is then integrated within each of these micro-cells. In this work, the recently developed ELI-D indicator was calculated and interpreted for the studied compounds. For

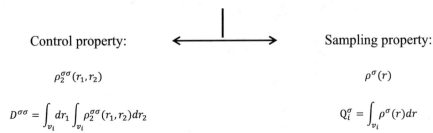

ELI-D restricted population approach

Control property: Sampling property:

$$\rho_2^{\sigma\sigma}(r_1, r_2)$$ $$\rho^{\sigma}(r)$$

$$D^{\sigma\sigma} = \int_{v_i} dr_1 \int_{v_i} \rho_2^{\sigma\sigma}(r_1, r_2) dr_2$$ $$Q_i^{\sigma} = \int_{v_i} \rho^{\sigma}(r) dr$$

Fig. 3.3 Scheme representing the restricted population approach used to define ELI-D

ELI-D the control property is the pair density, $\rho_2^{\sigma\sigma}(r_1, r_2)$ (see Sect. 3.1.2, Eq. 3.5) and the sampling property is the electron density, $\rho^{\sigma}(r)$, where σ is the spin, both α or β.

The charge Q_i^{σ} (sampling quantity) and the number of same spin electron pairs $D^{\sigma\sigma}$ (control quantity) are given as the respective integral of the electron density $\rho^{\sigma}(r)$ and the same spin pair density $\rho_2^{\sigma\sigma}(r_1, r_2)$ over the given micro-cell volume (v_i), as schematized in (Fig. 3.3).

After several mathematical/approximation steps [14] the ELI-D can be formulated as the product between electron density and the pair volume function $\tilde{V}_D^{\sigma}(r)$ (Eq. 3.15).

$$\tilde{\Upsilon}_D^{\sigma}(r) = \rho^{\sigma}(r) \cdot \tilde{V}_D^{\sigma}(r) \tag{3.19}$$

The pair volume function is defined as the volume needed to locally encompass a fixed fraction of a same-spin electron pair of σ-spin electrons, $D^{\sigma\sigma}$. Their product gives the charge per fixed number of same-spin pairs. Moreover, it was also demonstrated that ELI-D is directly proportional to the probability to find one electron inside a micro-cell and all the others outside. Thus, since an electron is localized if it repels same-spin pairs from its region, it is revealed by a high ELI-D value in that region. Finally, it is possible to conclude that ELI-D discerns regions with high electron localizability which means that the ELI-D topology is useful to highlight chemically sound regions such as core and valence shells, bonds and lone pairs, that correspond to ELI-D local maxima. ELI-D values in the core regions are typically in the range $0 \leqslant \tilde{\Upsilon}_D^{\sigma} \leqslant 10$ or more for heavier elements, whereas the valence regions display values within the range $0 \leqslant \tilde{\Upsilon}_D^{\sigma} \leqslant 2$. As an example, the ELI-D distribution for the CO molecule is shown in Fig. 3.4.

The topology of the ELI-D can be studied through its Laplacian [15]:

$$\nabla^2 \Upsilon_D^{\sigma} = \tilde{V}_D \cdot \nabla^2 \rho + \rho \cdot \nabla^2 \tilde{V}_D + 2\nabla\rho \cdot \nabla\tilde{V}_D \tag{3.20}$$

and its relative Laplacian:

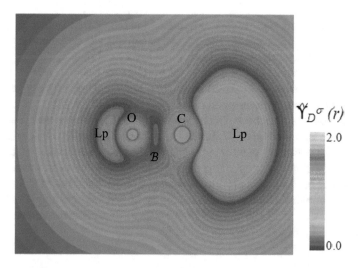

Fig. 3.4 ELI-D distribution for the CO molecule. Areas with high ELI-D values are: Lp = lone pairs; C = carbon core shell; O = oxygen core shell; $\mathcal{B}$ = C–O covalent bond

$$\frac{\nabla^2 \Upsilon_D^\sigma}{\Upsilon_D^\sigma} = \frac{\nabla^2 \rho}{\rho} + \frac{\nabla^2 \tilde{V}_D}{\tilde{V}_D} + 2\frac{\nabla \rho}{\rho} \cdot \frac{\nabla \tilde{V}_D}{\tilde{V}_D} \qquad (3.21)$$

Among them, $\nabla^2 \Upsilon_D^\sigma / \Upsilon_D^\sigma$ is particularly interesting since it can be expressed as a sum involving only the relative Laplacian of both electron density and pair volume function plus an additional mixed term thus, the ELI-D topology can be investigated on the basis of the topology of its constituent. In this way the similarity between the ELI-D and that of the $\nabla^2 \rho$ [10] can be rationalized. This position-space decomposition of the ELI-D was proven to be a useful tool for the study of chemical bonding for different systems allowing also to free from the strict statement that there is/is not a bond if an ELI-D attractor occurs/doesn't occur [15, 16].

Another important property of the ELI-D is that it can also be decomposed (charge decomposition) into additive positive partial ELI-D (pELI-D) contributions in the same way as electron density can be decomposed [14] for instance from orbitals or orbital groups within a chosen energy range or, in the case of solid state calculations, a region in k-space.

Since ELI-D is a scalar field, like $\rho(r)$, we can obtain an ELI-D space partitioning based on the ELI-D gradient vector field in terms of basins. From their location and connection to neighboring ELI-D basins they are classified as core and valence basins, the latter often being further subdivided into bond and lone pair [17, 18]. Integration of electron density within ELI-D basins yields their average electronic population, similar to Eq. (3.18). It is important to note that the electronic occupation of each atomic shell basin in many cases (but not all) resembles the number of electrons for each shell given by the Aufbau principle [19, 20]. If it is not the case, accurate quantitative studies based on the number of valence electrons cannot be conducted.

Since during this work some exceptions, like Zn and La, were encountered, an appropriate corrective method, described in the next paragraph, was developed and applied. The ELI-D valence basins are also interpreted in terms of their synaptic order [21]. This quantity is defined as the number of atomic core basins sharing a common surface with the considered valence basin $(\mathcal{B}_i)$. For simple main group molecules, the monosynaptic basins correspond to lone pair basins and the polysynaptic ones to chemical bonds. The idea that polycentric bonding can be related to the synapticity of valence basins seems to work well for molecules but not for solids: in general, polysinapticity is the necessary but not sufficient condition to have multicenter bonding in solids: it depends on the participation of each QTAIM atom to the considered ELI-D valence basins which need to be somewhat quantified. For this purpose, the ELI-D/QTAIM intersection technique is of great importance and allows to infer the effective atomicity of ELI-D valence basins which is the effective number of atom with which the considered basin is bonded to.

3.4   The ELI-D/QTAIM Intersection and the Position-Space Representation of the 8–N Rule[1]

The Bader mathematical formalism applied to electron density and ELI-D fields give two different complete and exhaustive partitioning of space in terms of the corresponding attractor basins. The superposition of both types of space partitioning within the ELI-D/QTAIM basin intersection method [22], yields on the one hand, which QTAIM atoms intersect a given ELI-D basin, i.e., a complete description of the ELI-D basin region in terms of atomic QTAIM regions. At the same time, it leads also to a complete description of the QTAIM atom in terms of regions of ELI-D basins. The inner part of a QTAIM atom is composed by ELI-D core basins whereas the outer part is composed by ELI-D valence basins, which are typically intersected by several QTAIM atomic basins. These basins make up the intersection set I_X of QTAIM atomic region Ω_X (atom X). After the intersection is performed, the next step is the determination of the electronic population inside the obtained basin segments, which can be utilized to define bond polarity by means of the bond fraction. The bond fraction of a specific ELI-D basin $\mathcal{B}_i$ (index i denotes all basins belonging to the unit cell) intersected by the QTAIM basin Ω_X of atom X is denoted as $p(\mathcal{B}_i^X)$. It is the ratio between the electronic populations of the intersected region $N(\mathcal{B}_i^X)$ and the total basin region $N(\mathcal{B}_i)$:

$$p(\mathcal{B}_i^X) = \frac{N(\mathcal{B}_i^X)}{N(\mathcal{B}_i)} \tag{3.22}$$

Hence, if $p(\mathcal{B}_i^X) = 1$ the ELI-D valence basins lies completely within the QTAIM atom X, representing a lone-pair of this atom, whereas, for a disynaptic basin a value

---

[1]The text of this section was originally published in [25].

equal to 0.5 indicate a non-polar covalent bond. Values between 0.5 and 1 indicate heteropolar bonds. The latter have been recently described [18, 23, 24] as composed by a nonpolar contribution, called the covalent part of the heteropolar bond, and a polar one, termed hidden lone-pair of the same bond. The non-polar covalent character $cc(\mathcal{B}_i)$ and polar (hidden) lone pair character $lpc(\mathcal{B}_i)$ of a basin $\mathcal{B}_i$ were calculated according to [25]:

$$cc(\mathcal{B}_i) = 2 - 2p(\mathcal{B}_i^X), \quad lpc(\mathcal{B}_i) = 2p(\mathcal{B}_i^X) - 1, \quad if \ p(\mathcal{B}_i^X) \geq 0.5 \qquad (3.23)$$

$$cc(\mathcal{B}_i) = 1, \quad lpc(\mathcal{B}_i) = 0, \quad if \ p(\mathcal{B}_i^X) < 0.5 \qquad (3.24)$$

$$with: \qquad cc(\mathcal{B}_i) + lpc(\mathcal{B}_i) = 1 \qquad (3.25)$$

The number of covalent $N_{cbe}(\mathcal{B}_i)$ and lone-pair $N_{lpe}(\mathcal{B}_i)$ electrons inside the X atom portion of bond basin $\mathcal{B}_i$ are evaluated according to:

$$N_{cbe}(\mathcal{B}_i) = N(\mathcal{B}_i) \cdot cc(\mathcal{B}_i) \qquad (3.26)$$

$$N_{lpe}(\mathcal{B}_i) = N(\mathcal{B}_i) \cdot lpc(\mathcal{B}_i) \qquad (3.27)$$

Then, a polar bond p can be decomposed into a covalent part b' and a hidden-lone-pair part lp' according to $p = [xb', (1-x)lp']$, where x denotes the fraction of two-electron covalent bond character being determined from the ELI-D/QTAIM basin intersection ($x = N_{cbe}(\mathcal{B}_i)/2$, see Eqs. 3.26/3.27). In order to obtain the corresponding atomic quantities, a summation of these electron contributions over the intersection set I_X of atom X has to be performed, which incorporates all basins $\mathcal{B}_i$ with a non-vanishing volume intersection with the QTAIM atomic region Ω_X. However, the intersection set I_X of atom X often contains ELI-D basins intersected at tiny portions, which are of no chemical relevance. In order to include only those basins of conceptual chemical relevance, solely those basins which touch the ELI-D core basin C^X of atom X were effectively included in the summations of Eq. (3.28/3.29). This set of basins has been called the access basin set s_X of atom X [23]. The access set is expected to be a non-proper subset of the intersection set of basin X, $s_X \subseteq I_X$.

$$N_{cbe}(X) = \frac{1}{2} \sum_{i=1}^{I_X} N_{cbe}(\mathcal{B}_i) \approx \frac{1}{2} \sum_{i=1}^{s_X} N_{cbe}(\mathcal{B}_i) \qquad (3.28)$$

$$N_{lpe}(X) = \sum_{i=1}^{I_X} N_{lpe}(\mathcal{B}_i) \approx \sum_{i=1}^{s_X} N_{lpe}(\mathcal{B}_i) \qquad (3.29)$$

The number of valence electrons of an atom X is exactly calculated as the difference of the electronic population of the QTAIM atom and the total core population according to ELI-D.

$$N_{val}^{ELI}(X) = N(\Omega_X) - N(\mathcal{C}^X) = \sum_{i=1}^{l_X} N(\mathcal{B}_i^X) \approx \sum_{i=1}^{s_X} N(\mathcal{B}_i^X) \qquad (3.30)$$

Within the ELI-D topological framework, the number of electrons an atom X has access to, its access electron number, was defined [23] as follow:

$$N_{acc}^{ELI}(\mathcal{C}^X) = \sum_{i=1}^{s_X} N(\mathcal{B}_i) \qquad (3.31)$$

For a number of binary and ternary intermetallic phases tested, the access electron number of main group atoms X (at the right of the Zintl border) has been shown to approximately obey relation Eq. (3.32) (with $N_{cb}(X) = \frac{1}{2}N_{cbe}(X)$).

$$N_{cb}(X) = N_{acc}^{ELI}(\mathcal{C}^X) - N_{val}^{ELI}(X) \qquad (3.32)$$

If the access electron number of the main group element species X is close to 8, Eq. (3.32) can be interpreted as a position space variant of the 8–N rule for atom X [23]. The definition of a chemically meaningful access electron number for this equation was the main reason for working with the access set instead of the intersection set. Hence, this approach extended the application field of the 8-N rule also to compounds composed by heteropolar polyanionic network for instance the MgAgAs-type phases, the so called half-Heusler phases and some zinc-blende-type compounds [18, 23, 24]. In some reported cases a significant mismatch between the Aufbau and the ELI-D shell population lead to $N_{acc}^{ELI}(\mathcal{C}^X)$ far from the 8 value with the results that the correspondence between Eq. (3.32) and the 8-N rule is somewhat hidden. This finding was reported for some binary compounds (*i.e.* GaAs, NaP, Na_2S_2) in [23]. The necessity to correct the valence electron count was indicated in [24] but it was not shown how to numerically apply such correction, in that case to the [yp; $(4-y)lp$] bond classification, *e.g.* for the discussed InSb and GaSe. During the thesis work the extension of this approach to include explicit adjustment of the valence electron count in position space has turned out to be necessary, in particular for the quantitative analysis of the chemical bonding between Ge based polyanionic networks and the surrounding metal atoms within the R_2MGe_6 compounds ($R =$ La, Y; $M =$ Li, Mg, Zn, Cu, Ag, Pd [25]—see Chap. 4).

3.5 Improved Methods for the Chemical Bonding Analysis of the R_2MGe_6 Intermetallics[2]

Compared to the previously reported procedure [18, 23, 24], Eq. (3.24) marks a slight deviation [25]. The reason behind is based on the observation, that due to small participation (small bond fractions $p(\mathcal{B}_i^X) \ll 0.5$) of metal atoms in Ge–Ge bond basins, bond fractions $p(\mathcal{B}_i^{Ge})$ of slightly below 0.5 were often found. The originally used formulae for the covalent and lone pair character of the majority owner X (Eq. 3.22) were based on the assumption that the corresponding bond fractions $p(\mathcal{B}_i^X)$ always obey $p(\mathcal{B}_i^X) \geq 0.5$. $p(\mathcal{B}_i^X)$ values slightly below 0.5 yields to covalent characters lightly above 1.0 and small but negative lone pair characters, such that $cc(\mathcal{B}_i) + lpc(\mathcal{B}_i) = 1$ is still obeyed. With Eq. (3.24) applied for $p(\mathcal{B}_i^X) <$ 0.5, negative lone pair characters are avoided. As an example, for $p(\mathcal{B}_i^{Ge}) = 0.48$, Eq. (3.23) would yield to $cc(\mathcal{B}_i) = 1.04$ and $lpc(\mathcal{B}_i) = -0.04$, whereas now with Eq. (3.34) $cc(\mathcal{B}_i) = 1$ and $lpc(\mathcal{B}_i) = 0$. It is to be noted, that while Eq. (3.33) holds for all cases $p(\mathcal{B}_i^X) \gtrless 0.5$,

$$N_{val}^{ELI}(X) = N_{cbe}(X) + N_{lpe}(X) \tag{3.33}$$

the access electron number was always calculated according to Eq. (3.31), because Eq. (3.34) holds only for $p(\mathcal{B}_i^X) \geq 0.5$ throughout the access set s_X of atom X.

$$N_{acc}^{ELI}(\mathcal{C}^X) = 2N_{cbe}(X) + N_{lpe}(X), \quad for\ all\ \mathcal{B}_i^X \in s_X : p(\mathcal{B}_i^X) \geq 0.5 \tag{3.34}$$

For $p(\mathcal{B}_i^X) < 0.5$ Eq. (3.34) is no longer strictly valid, and may only be approximately fulfilled depending on the amount of deviation from 0.5. In this case, also the position-space 8–N type of formula Eq. (3.32) may only be approximately fulfilled. For the extreme case of, *e.g.* three-atomic character of the basin intersections, the formalism has to be suitably extended leading to modifications of Eq. (3.32), which has been considered as a future perspective [24].

With the knowledge of the atomic quantities $N_{val}^{ELI}(X)$ and $N_{acc}^{ELI}(\mathcal{C}^X)$, the genuine charge claim $P_X(X)$ has been introduced to characterize the atomic species X. If all bond fractions of the access set $p(\mathcal{B}_i^X) \geq 0.5$ is valid, it can be considered as an average bond charge of all basins in the access set s_X of atom X and adopts values in the range [0; 1] like the bond fraction Eq. (3.22).

$$P_X(X) = \frac{\sum_{i=1}^{s_X} p(\mathcal{B}_i^X)N(\mathcal{B}_i)}{\sum_{i=1}^{s_X} N(\mathcal{B}_i)} = \frac{\sum_{i=1}^{s_X} N(\mathcal{B}_i^X)}{N_{acc}^{ELI}(\mathcal{C}^X)} \approx \frac{N_{val}^{ELI}(X)}{N_{acc}^{ELI}(\mathcal{C}^X)},$$
$$for\ all\ \mathcal{B}_i^X \in s_X : p(\mathcal{B}_i^X) \geq 0.5 \tag{3.35}$$

---

[2]The text of this section was originally published in [25].

The number of covalent bonding and lone pair electrons of an atomic species X can then be obtained from the genuine charge claim in analogy to Eqs. (3.23/3.24) and (3.28/3.29) according to:

$$N_{cbe}(X) = \frac{1}{2} N_{acc}^{ELI}\left(C^X\right) \cdot (2 - 2P_X(X)), \quad \mathcal{B}_i^X \in s_X : p\left(\mathcal{B}_i^X\right) \geq 0.5 \qquad (3.36)$$

$$N_{lpe}(X) = N_{acc}^{ELI}\left(C^X\right) \cdot (2P_X(X) - 1), \quad \mathcal{B}_i^X \in s_X : p\left(\mathcal{B}_i^X\right) \geq 0.5 \qquad (3.37)$$

Since this primary condition is not well fulfilled for most of the Ge species, usage of the genuine charge claim has been skipped in the chemical bonding discussion reported in Chap. 4.

ELI-D is known to reveal atomic shell structure in an approximately quantitative way [19, 20]. This means that deviations between the integer values obtained from the Aufbau principle and those obtained from integration of the electron density between the ELI-D shell basin boundaries typically display values of a few tenths of an electron. As already mentioned, this has not been accounted for in previous work. Now, the incorporation of these corrections will be shown to be important to adequately characterize different bonding interactions like the La–Ge and the Zn–Ge ones. A schematic sketch of the whole procedure is given in Fig. 3.5.

The first step is the correction of the penultimate shell's electronic population with respect to the inner shells and an assumed number of valence electrons. If the number of chemically active valence electrons $N_{val}^{chem}(X)$ is known, the number of

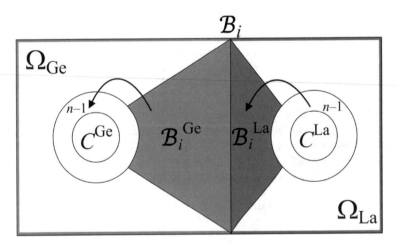

Fig. 3.5 Schematic representation of the PSC0 correction for a diatomic La–Ge ELI-D bond basin $\mathcal{B}_i$. The encompassing rectangular regions Ω_{Ge} (orange lines) and Ω_{La} (green lines) represent the QTAIM atoms Ge and La, respectively. They cut bond basin region $\mathcal{B}_i$ into a part $\mathcal{B}_i^{Ge}$ (filled orange) and $\mathcal{B}_i^{La}$ (filled green). Spherical regions "n–1" denote the spatial regions of the n–1 atomic shells of Ge and La with n being the main quantum number of the atoms valence shells. Reproduced with permission from [25]

chemically inactive core electrons up to the $n-1$ shell $N^{chem}_{n-1}(X)$ (n denotes the atom's period number in the periodic table of the elements PTE) becomes fixed.

$$N^{chem}_{n-1}(X) = N^{PTE}(X) - N^{chem}_{val}(X) \tag{3.38}$$

The number of ELI-D core or valence shell defects, $N^{ELI}_{cdef}(X)$ or $N^{ELI}_{vdef}(X)$, respectively, is given by simply adding all ELI-D core shell populations including the penultimate shell and comparison with the value of $N^{chem}_{n-1}(X)$.

$$N^{ELI}_{cdef}(X) = N^{chem}_{n-1}(X) - \sum_{i=1}^{n-1} N^{ELI}_i(X) = -N^{ELI}_{vdef}(X) \tag{3.39}$$

Corresponding core corrections to the valence electron count are appropriate for all atom types occurring in the title compounds. For example, from atomic shell structure investigation of free atoms [20] Ge displays a core underpopulation of $N^{ELI}_{cdef}(Ge)$ $= -0.265$ e, $i.e.$ a valence shell overpopulation of $N^{ELI}_{vdef}(Ge) = +0.265$ e. For the Ge species in each of the compounds analyzed, very similar values are obtained. In order to reassign in the second step the surplus valence electrons from the valence basins attached to Ge atoms back to the core region, a spherical approximation has been chosen, denoted as PSC0 (penultimate shell correction of lowest order) in the following. The outer side of the Ge penultimate shell has a total surface area of $A_{n-1}(Ge)$. It touches all valence basins of the access set s_{Ge}. This way each such valence basin displays a common surface ("patch") with area $A(\mathcal{B}_i)$ with the outer side of the Ge penultimate shell. The patch areas have been determined using a built-in functionality of the graphical program AVIZO [26]. The electronic population of each of these basins $\mathcal{B}_i$ is corrected by an amount proportional to the ratio of the partial and the total surface area ($X = Ge$, Eq. 3.40),

$$N_{corr}(\mathcal{B}_i) = -N^{ELI}_{vdef}(X) \cdot \frac{A(\mathcal{B}_i)}{A_{n-1}(X)} \tag{3.40}$$

where the partial surface areas ("patch areas") sum up to the total one,

$$A(X) = \sum_{i=1}^{s_X} A(\mathcal{B}_i) \tag{3.41}$$

It is now important to realize that the correction value $N_{corr}(\mathcal{B}_i)$ is used to directly modify the intersection value $N(\mathcal{B}^X_i)$, which indirectly modifies also the total population of the basin $\mathcal{B}_i$ by the same amount. This way, the $X = Ge$ atomic valence population becomes modified. As a result, the corrected bond fraction (cf. Eq. 2.22) for Ge reads

$$p_{corr1}\left(\mathcal{B}^{Ge}_i\right) = \frac{N\left(\mathcal{B}^{Ge}_i\right) + N_{corr1}(\mathcal{B}_i)}{N(\mathcal{B}_i) + N_{corr1}(\mathcal{B}_i)} \tag{3.42}$$

In addition, all the metal atoms (besides Zn) display valence shell under-populations, *e.g.*, $N_{vdef}^{ELI}(\text{Mg}) = -0.0884$. Again, the population of each basin $\mathcal{B}_i$ touching the Mg penultimate shell is corrected by an amount calculated according to Eq. (3.40). However, this value is only added to the total basin population and is not included, in the Ge population part of the bond fraction, *i.e.* it does not change the Ge atomic valence population $N_{val}^{ELI}(\text{Ge})$. The bond fraction for this correction reads

$$p_{corr2}\left(\mathcal{B}_i^{\text{Ge}}\right) = \frac{N\left(\mathcal{B}_i^{\text{Ge}}\right)}{N(\mathcal{B}_i) + N_{corr2}(\mathcal{B}_i)} \tag{3.43}$$

Both types of corrections are to be performed simultaneously,

$$p_{corr}\left(\mathcal{B}_i^{\text{Ge}}\right) = \frac{N\left(\mathcal{B}_i^{\text{Ge}}\right) + N_{corr1}(\mathcal{B}_i)}{N(\mathcal{B}_i) + N_{corr1}(\mathcal{B}_i) + \sum_{i=1}^{s_{\mathcal{B}_i}} N_{corr2,i}(\mathcal{B}_i)} \tag{3.44}$$

where s_{B_i} denotes the access set of basin $\mathcal{B}_i$, those core basins with which it has a common surface. Note that, if basin $\mathcal{B}_i$ is a Ge–Ge' bond basin, the bond fraction $p_{corr}(\mathcal{B}_i^{\text{Ge}})$ contains the corr1 type of correction only once (for Ge), the attached Ge' species only enters here in the *corr2* type of correction.

The chosen number of valence electrons for the different species in the title compounds is Li (1 ve), Mg (2 ve), Al (3 ve), Ge (4 ve), Zn (2 ve), and La (3 ve). The spreadsheets generated to apply the PSC0 correction to La_2MgGe_6 are exemplarily shown in Tables A3.1–A3.8.

For atomic species $M = \text{Cu}$, Ag, and Pd no simple guess of active valence electrons is possible, which makes the corresponding compounds $\text{La}_2 M \text{Ge}_6$ unsuitable for calculation of bond polarities with the actual methodology.

Another important classical chemical quantity is the oxidation number calculated assigning all the valence electrons to the more electronegative elements. In the same way, on the basis of the bond fractions (Eq. 3.22), the conceptual ELI-D based oxidation numbers (ELIBON) [27] were defined. More precisely, for an atom X with the s_X access set and core basin C^X, the ELIbond is calculated assigning the whole bond basins population $N(\mathcal{B}_i)$ to the atom being the majority owner of the electronic population, according to the QTAIM intersection [17, 28]. Basin populations between same types of majority atoms are equally distributed between them. When the abovementioned PSC0 correction is applied, the ELIbon values for metal atoms (which are never the major owner of any valence basins) exactly correspond to the chosen number of valence electrons (*i.e.* +1 for Li, +2 for Mg, etc....). On the other hand, the ELIbons obtained for different Ge species are not predetermined, and it is interesting to compare them with the formal ones.

3.6 Software Employed

Density Functional Theory (*DFT*) calculations were carried out with different software depending on the targeted investigations for selected compounds. The plane wave pseudopotential code QUANTUM-ESPRESSO [29] was used to evaluate total energies of different polymorphs, with R_2MGe_6 general formula, related to the same aristotype through group-subgroup relations. For position-space chemical bonding analysis of the just mentioned phases, preliminary *DFT* calculations were performed by means of FPLO [30] (Full-Potential Local-Orbital) and FHI-aims [31] (Fritz Haber Institute ab initio molecular simulations) codes. At the end of the self-consistent field calculation the FPLO package gives, thanks to a Dresden MPI-CPfS ChemBond group implementation, the electron density and the ELI-D numerically calculated in a regular grid. Then, real space analysis like basin determination and integration, are performed on the basis of the obtained ρ and ELI-D by means of the software DGrid-5.0 [32]. On the other hand, FHI-aims gives direct access to the wave function which is the input function employed by DGrid-5.0 to get analytically ρ, ELI-D and also other fields, like their Laplacians, not accessible with FPLO results. Bond fractions, number of covalent and lone-pair electrons, and the position space variant of the 8-*N* rule were calculated/applied with program BondFraction [33]. Scalar field topologies in the direct space and the associated basins were drawn with graphical program AVIZO [26] and Paraview [34].

The *DFT*-based TB-LMTO-ASA (Linear Muffin Thin Orbitals-Atomic Sphere Approximation) [35] package was used to obtain COHP curves and integrated COHP values (*i*COHP), further visualized with WxDragon [36].

More details about the set-up of the calculations (number of *k*-points, exchange and correlation functional, etc....) are described in the following chapters.

References

1. Oganov AR (2010) Modern methods of crystal structure prediction. Wi e -V H Verlag GmbH & Co. KGaA
2. Steinberg S, Dronskowski R (2018) Crystals 8:225. https://doi.org/10.3390/cryst8050225
3. Dronskowski R, Blöchl PE (1993) J Phys Chem 97:8617–8624
4. Bende D, Wagner FR, Sichevych O, Grin Yu (2017) Angew Chem Int Ed 56:1313–1318
5. Hohenberg P, Kohn W (1964) Phys Rev 136:B864
6. Matta CF, Gillespie RJ (2002) J Chem Ed. 9:1141–1152
7. Schrödinger E (1926) Ann Phys 384:361–376
8. Khon W, Sham LJ (1965) Phys Rev 140:A1133
9. Bloch F (1929) Z Physik 52:555–600
10. Bader RFW (1990) Atoms in molecules: a quantum theory. Oxford University Press, Oxford
11. Guillemin V, Pollack A (2010) Differential topology. AMS Chelsea Publishing
12. Kohout M (2004) Int J Quantum Chem 97:651–658
13. Pendas AM, Kohout M, Blanco MA, Francisco E (2012) In: Gatti C, Macchi P (eds) Modern charge-denstity analysis. Springer, Heidelberg, London, New York, pp 303–358
14. Wagner FR, Bezugly V, Kohout M, Grin Y (2007) Chem Eur J 13:5724–5741

46   3   Theoretical Methods

15. Wagner FR, Kohout M, Grin Y (2008) J Phys Chem 112:9814–9828
16. Wagner FR, Cardoso-Gil R, Boucher B, Wagner-Reetz M, Sichelschmidt J, Gille P, Baenitz M, Grin Y (2018) Inorg Chem 57:12908–12919
17. Kohout M, Pernal K, Wagner FR, Grin Y (2004) Theor Chem Acc 112:453–459
18. Bende D, Grin Y, Wagner R (2016) In: Felser C, Hirohata A (eds) Heusler alloys. Springer International Publishing, Switzerland pp 133–155
19. Kohout M, Savin A (1996) Int J Q Chem 60:875–882
20. Baranov AI (2014) J Comput Chem 35:565–585
21. Savin A, Silvi B (1996) Colonna, F. Canadian J Chem-Revue Canadienne de Chimie 74:1088–1096
22. Butovskii MV, Tok OL, Wagner FR, Kempe R (2008) Angew Chem Int Ed 47:6469–6472
23. Bende D, Wagner FR, Grin Y (2015) Inorg Chem 54:3970–3978
24. Wagner FR, Bende D, Grin Y (2016) Dalton Trans 45:3236–3243
25. Freccero R, Solokha P, De Negri S, Saccone A, Grin Y, Wagner FR (2019) Polar-covalent bonding beyond the Zintl picture in intermetallic rare earth germanides. Chem Eur J 25:6600–6612. https://doi.org/10.1002/chem.201900510 Copyright Wiley-VCH Verlag GmbH & Co. KGaA. Reproduced with permission
26. Program AVIZO, version 9.1, Fisher Scientific
27. Veremchuk I, Mori T, Prots Yu, Schnelle W, Leithe-Jasper A, Kohout M, Grin Y (2008) J Solid State Chem 181:1983–1991
28. Höhn P, Agrestini S, Baranov A, Hoffmann S, Kohout M, Nitsche F, Wagner FR, Kniep R (2011) Chem Eur J 17:3347–3351
29. Giannozzi P, Baroni S, Bonini N, Calandra M, Car R, Cavazzoni C, Ceresoli D, Chiarotti GL, Cococcioni M, Dabo I, Dal Corso A, de Gironcoli S, Fabris S, Fratesi G, Gebauer R, Gerstmann U, Gougoussis C, Kokalj A, Lazzeri M, Martin-Samos L, Marzari N, Mauri F, Mazzarello R, Paolini S, Pasquarello A, Paulatto L, Sbraccia C, Scandolo S, Sclauzero G, Seitsonen AP, Smogunov A, Umari P, Wentzcovitch RM (2009) J Phys Condens Matter 21:395502
30. Koepernik K, Eschrig H (1999) Phys Rev B 59:1743
31. Blum V, Gehrke R, Hanke F, Havu P, Havu V, Ren X, Reuter K, Scheffler M (2009) Comput Phys Commun 180:2175–2196
32. Kohout M (2017) Program *Dgrid*, version 5.0. Radebeul
33. Wagner FR (2017) Program BondFraction, version 5. Max-Planck-Institut für Chemische Physik fester Stoffe, Dresden, Germany
34. (a) Ayachit U (2015) The paraview guide: a parallel visualization application. Kitware. ISBN 978-1930934306; (b) Baranov AI (2015) Visualization plugin for paraview. version 1.3
35. Krier G, Jepsen O, Burkhardt A, Andersen OK (2000) The TB-LMTO-ASA Program. Version 4.7; Max-Planck-Institut für Festkörperforschung, Stuttgart, Germany
36. Eck B (1994-2018) Wxdragon. Aachen, Germany. Available at http://wxdragon.de

Chapter 4
The R_2MGe_6 ($R = $ Rare Earth Metal; M $= $ Another Metal) Compounds: An Experimental and Theoretical Study on Their Synthesis, Crystal Structure and Chemical Bonding

Germanium–rich compounds of general formula R_2MGe_6 ($R =$ rare earth metal, $M =$ another metal) form a numerous family, which has been investigated since the eighties of the last century up to the present [1]. The scientific interest in these phases arises from the fact that they exist with many different metals M, ranging from the s–block ($M = $ Li, Mg) to the p–block ($M = $ Al, Ga) of the periodic table, including many transition elements. That is why, these compounds are well suited for systematic studies on crystal structure–electronic structure–property relationships, aiming to evidence for example the role of both R and M properties (size, electronegativity, valence electrons, etc…) in the chemical bonding scenario, set up by characteristic Ge–based covalent fragments.

A variety of physical properties, including electrical resistivity, magnetic susceptibility, specific heat and thermoelectric power, have already been measured on several R_2MGe_6 representatives, mostly with transition elements, such as Co, Ni, Pd, Pt, Cu, as M components [2–8]. Many of the investigated compounds exhibit antiferromagnetic ordering below ~ 30 K [2, 3, 7]; for some Yb$_2MGe_6$ phases a behaviour characteristic of intermediate valence systems is reported [4, 8, 9].

Despite the great amount of experimental work, controversial data exist both on the interpretation of physical/magnetic properties and on the crystal structures of the R_2MGe_6 compounds. They are reported to crystallize in orthorhombic or monoclinic space groups (see Table 4.1), distributing among the following prototypes: $oS18$–Ce$_2$CuGe$_6$ (space group: $Amm2$), which is the most represented, $oS72$–Ce$_2$(Ga$_{0.1}$Ge$_{0.9}$)$_7$ (space group: $Cmce$) and $mS36$–La$_2$AlGe$_6$ ($C2/m$). A different orthorhombic structure (space group: $Cmmm$) has been assigned to a few R_2LiGe$_6$ ($R = $ La, Ce, Pr) phases, referred to the $oS18$–Pr$_2$LiGe$_6$ prototype [10–12] and to Yb$_2$Pd$_{1.075(1)}$Ge$_6$, referred to the $oS20$–SmNiGe$_3$ prototype [13].

The $oS72$–Ce$_2$(Ga$_{0.1}$Ge$_{0.9}$)$_7$ structure was found to be the correct one for a number of compounds previously reported as $oS18$, such as La$_2$PdGe$_6$, Dy$_2$PdGe$_6$ [6, 15]

© The Editor(s) (if applicable) and The Author(s), under exclusive license to Springer Nature Switzerland AG 2020
R. Freccero, *Study of New Ternary Rare-Earth Intermetallic Germanides with Polar Covalent Bonding*, Springer Theses, https://doi.org/10.1007/978-3-030-58992-9_4

Table 4.1 Structure types of R–M–Ge compounds of ~ 2:1:6 stoichiometry ($R = $ rare earth metal; $M = $ another metal). A blank cell means that an alloy of this composition was not investigated

Group	M	R= Y	La	Ce	Pr	Nd	Sm	Eu	Gd	Tb	Dy	Ho	Er	Tm	Yb	Lu
I	Li		▲	▲	▲											
II	Mg	−	□	−	−	−	−	−	−	−	−	−	−	−	−	−
VII	Mn		⊠	⊠												
IX	Co	⊠							⊠	⊠	⊠	⊠	⊠	⊠		⊠
	Rh						⊠									
X	Ni	⊠			⊠	⊠	⊠	−	⊠	⊠	⊠	⊠	⊠	⊠	⊠	⊠◇
	Pd	⊠	⊠□	⊠	⊠	⊠	⊠		⊠	⊠	⊠□	⊠	⊠	⊠	⊠□	
	Pt	⊠			⊠	⊠	⊠	⊠	⊠	⊠	⊠	⊠	⊠	⊠	⊠	
XI	Cu	⊠	⊠	⊠	⊠	⊠	⊠	−	⊠	⊠	⊠	⊠	⊠	⊠	⊠◇	
	Ag		⊠	⊠	⊠	⊠	⊠	−	⊠			−			−	
	Au			⊠	⊠	⊠	⊠	−	⊠	⊠	⊠	−				
XII	Zn	◆	□	□	□	□	□		□	□	◆	◆	−	−		
XIII	Al	−	◇	◇	◇	◇	◇	−	◇	◇	⊠	−	−	−	−	−
	Ga	□◇		□				−								

⊠ = $oS18$–Ce$_2$CuGe$_6$ SG: *Amm2* (№ 38)   □ = $oS72$–Ce$_2$(Ga$_{0.1}$Ge$_{0.9}$)$_7$ SG: *Cmce* (№ 64)

◇ = $mS36$–La$_2$AlGe$_6$ SG: *C2/m* (№ 12)   ▲ = $oS18$–Pr$_2$LiGe$_6$ SG: *Cmmm* (№ 65)

◆ = $mP34$–Dy$_2$Zn$_{1-x}$Ge$_6$ SG: *P2/m* (№ 11)   − = not observed

Adapted with permission from [14]

and Yb$_2$PdGe$_6$ [8, 16], and for several new 2:1:6 representatives, such as La$_2$MgGe$_6$ [17] and R_2ZnGe$_6$ [14]. Recently, the alternative synthesis in metal flux was applied to Yb$_2$CuGe$_6$, and the $mS36$–La$_2$AlGe$_6$ crystal structure was proposed for this compound, instead of the previously reported $oS18$–Ce$_2$CuGe$_6$ type [18]. In addition, it is also important to highlight that the compositions of many of the abovementioned R–M–Ge phases are not exactly coincident with the 2:1:6 stoichiometry, being Ge or M richer. It is the case of two prototypes: in Ce$_2$(Ga$_{0.1}$Ge$_{0.9}$)$_7$ a statistical mixture of Ga and Ge in different crystallographic sites leads to a Ge–richer composition, instead in La$_2$AlGe$_6$ (La$_2$Al$_{1.6}$Ge$_{5.4}$) a partial substitution of Ge by Al atoms is reflected in a Ge–poorer composition. In these cases, the total number of atoms per unit cell is coincident with the stoichiometric model, but a more correct general formula should be R_2M_{1-x}Ge$_{6+x}$, where x could be positive or negative. For $R = $ Y, $M = $ Ga both possibilities are realized [19]; in fact two compounds have been reported with these elements, a Ge–rich phase (Y$_2$Ga$_{0.34}$Ge$_{6.66}$, $oS72$, x = 0.66) and a Ge–poor one (Y$_2$Ga$_3$Ge$_4$, $mS36$, x = −2). Another interesting behavior was found along the Zn–containing series of compounds, where the crystal structures change from the orthorhombic $oS72$–Ce$_2$(Ga$_{0.1}$Ge$_{0.9}$)$_7$ structure ($R = $ La–Nd, Sm, Gd, Tb) to the new monoclinic structure $mP34$–Dy$_2$Zn$_{1-x}$Ge$_6$ (x ~ 0.5, $R = $ Dy, Ho, Y) was observed. The latter is an ordered superstructure of the La$_2$AlGe$_6$ prototype.

Structural relations can be found between some of the aforementioned modifications. Invoking the linear intergrowth concept [20], the $oS72$ and $mS36$ can be viewed as belonging to a homological series constructed by linear intergrowth of

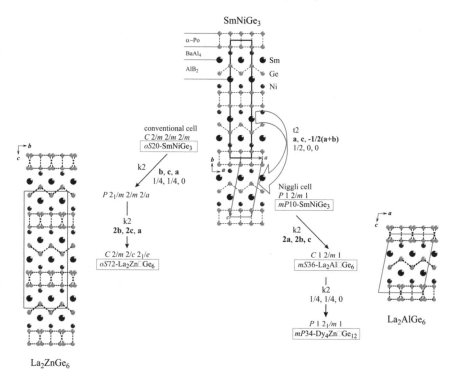

Fig. 4.1 Bärnighausen tree relating the SmNiGe$_3$ aristotype and its orthorhombic ($oS72$) and monoclinic ($mS36$ and $mP34$) vacancy variants. The type and indexes of the symmetry reductions are given. Reprinted with permission from [14]

inhomogeneous segments of the defective BaAl$_4$, AlB$_2$ and α–Po structure types [14, 21]. An alternative description was given by Grin [20], based on only two types of segments, AlB$_2$ and defective CeRe$_4$Si$_2$. A more generalized scheme, based on the vacancy ordering criterion within the group–subgroup theory, was proposed to describe relations between the $oS72$, $mS36$ and $mP34$ modifications [14]. The resulting scheme (Fig. 4.1) is a two–branched Bärnighausen tree originating from the SmNiGe$_3$ aristotype via subsequent reduction steps accompanied by vacancy ordering.

Within this tree, the 2:1:6 compounds with $oS72$ and $mS36$ structures can be viewed as "isomers" located each on a different branch, both obtained after two symmetry reductions steps. Growing longer the monoclinic branch from the $mS36$ model, the $mP34$ structure (corresponding to the compound Dy$_2$Zn$_{1-x}$Ge$_6$ $\approx$ Dy$_4$ZnGe$_{12}$) is obtained via one more reduction step along with further vacancy ordering.

Being strictly related, these phases originate very similar X–ray powder pattern making almost impossible to assign the correct model when dealing with multiphase samples. It is also the case for the $oS18$–Ce$_2$CuGe$_6$, as shown in Fig. 4.2.

Fig. 4.2 Comparison between calculated X–ray powder patterns in the range between 15° and 45° for Pr$_2$PdGe$_6$ with $oS72$, $mS36$ and $oS18$ structure. Crystal data of the first two modifications were found during this work whereas the last one is from the literature

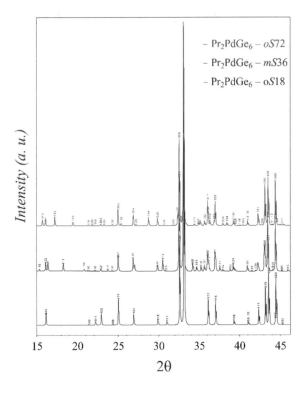

As it is clear from the similarity between calculated patterns shown in Fig. 4.2, a structural relation should exist also with the most reported $oS18$ modification and its identification was one of the aim of this work. In addition, the diffraction patterns shown evidence that powder X–ray diffraction data could be sufficient to assign the correct model only in the case of high quality patterns of almost single phase samples, generally quite difficult to obtain. For these reasons structural determination/revision in this work were only performed on the basis of single crystal X–ray data. The fact that the $oS18$–Ce$_2$CuGe$_6$ model was almost always deduced by means of the X–ray powder diffraction technique and that some of the aforementioned experimental results revealed the $oS72$ and $mS36$ to be the correct ones, indicates that the most represented structural type for the R_2MGe$_6$ phases should be probably revised.

In fact, a correct structural description would be of fundamental importance not only to appropriately interpret results of physical properties characterizations but also to evidence differences, similarities and regularities among the numerous 2:1:6 compounds in view of further chemical bonding studies.

The bonding in R_2MGe$_6$ can be initially addressed by the Zintl–Klemm concept [22, 23]. Even though, on the basis of interatomic distances it is reasonable to suggest the presence of two–bonded Ge ($2b$) zigzag chains and three–bonded Ge ($3b$) corrugated layers for all compounds, the Zintl–Klemm concept can be successfully applied only when M is a divalent metal, like Mg and Zn. In that cases the simplified ionic

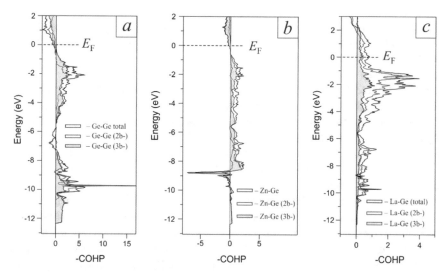

Fig. 4.3 COHP curves for the La_2ZnGe_6 intermetallic: **a** Ge–Ge, **b** Zn–Ge and **c** La–Ge interactions. Reprinted with permission from [14]

formula $(R^{3+})_2(M^{2+})[(2b)Ge^{2-}]_2[(3b)Ge^-]_4$ leads to electroneutrality. Even if the Zintl–Klemm approach is not fulfilled when M is not divalent, it does not prevent the studied compounds to have closely related structures with the same Ge covalent fragments. Thus, it is clear that there is more to understand about the chemical bonding for title phases, suggesting to overcome the approximation of a complete charge transfer from metals to Ge species. Some thorough studies, conducted on the basis of DOS and COHP analysis [14, 24–27], revealed the importance of considering partial charge transfer. This yields bonding interactions between cationic and anionic partial structures, which should be polar–covalent, instead of ionic, and are suspected to feature multicentric character [27]. They have been made responsible for metallic properties in representatives, which even formally obey the 8–N rule. For instance, it is the case for the nominally electron–precise AE_5Ge_3 (AE = Ca, Sr, Ba) phases [24]. The non–zero density of states at the Fermi level was assign to a significant mixing between Ca d– and Ge p–states. Among the 2:1:6 stoichiometry phases, electronic structure calculations were performed only for La_2ZnGe_6 and chemical bonding investigated through DOS/COHP curves (Fig. 4.3) [14].

The presence of infinite $(2b)$Ge zigzag chains and $(3b)$Ge corrugated layers was confirmed and the bonding relevance of Ge–Zn and Ge–La interactions shown.

In order to give more insight into chemical bonding for the title phases, the position–space analysis was applied. In fact, this approach based on quantum chemical topology analysis of the electron density (QTAIM method [28]) and ELI–D/ELF scalar fields was already successfully applied to numerous binary and ternary intermetallic compounds including borides [29, 30], carbides [31] and Ge–based clathrates [32]. Recently a quantum chemical position–space representation and extension of the classical 8–N rule was introduced and applied to binary Zintl compounds and Half–Heusler ternary phases [33, 34]. While these latter studies were focused on polar bonding within the polyanionic network, the present bonding analysis focused on the complementary part, the analysis of polar bonding between the polyanionic Ge network and the surrounding metal atoms and also to metal–metal bonding. For the description of polar network–metal bonding in the title compounds, the previous approach [33, 34] was extended in order to obtain a consistent and balanced picture. Moreover, the presence of metal atoms coordinatively unsaturated by the polyanionic network, gave rise to various degrees of metal–metal interactions. This issue will be discussed employing also the novel fine structure analysis of the ELI–D distribution, which finally allowed to recover the detailed bonding scenarios and their gradual changes.

In this chapter a complete structure revision of the R_2PdGe_6, together with the synthesis of some new representatives, is reported. The Pd–containing intermetallics were chosen since they are reported to exist for many rare earth metals and that the structures of some of them were recently revised by single crystal experiments. In this framework, it was also decided to investigate the R_2LiGe_6 series of compounds, which are, surprisingly, the only one reported to crystalize with a $oS18$–Pr_2LiGe_6 structure type. The challenging experimental work necessary for this achievement was also complemented by DFT total energy calculations, in order to generalize the obtained results to the whole R_2MGe_6 family and guide future investigations on new representatives. In order to experimentally check the correctness of the revision for all R_2MGe_6, La_2CuGe_6 and La_2AgGe_6 were prepared and characterized. Since correct crystal structure are the necessary starting point for a meaningful chemical bonding study, obtained results allowed to perform a comprehensive chemical bonding analysis over the whole family. The orthorhombic $oS72$ compounds La_2MgGe_6 [17], La_2ZnGe_6 [14] and Y_2PdGe_6 [35], the monoclinic $mS36$ La_2AlGe_6 [21] and La_2PdGe_6 [35] and La_2LiGe_6, La_2CuGe_6 and La_2AgGe_6, were selected for the real space chemical bonding analysis [36]. The compounds chosen display varying total electron counts encompassing also the ideal 8–N one. This way, stepwise changes of the bonding scenario are expected to occur, which are to be investigated in detail.

4.1 Synthesis and Phase Analysis[1]

Different synthetic routes were followed, all starting from the pure components.

Samples of about 0.8 g with $R_{22.2}Pd_{11.1}Ge_{66.7}$ (R = Sc; Y; La–Nd; Sm–Lu), $R_{22.2}Li_{11.1}Ge_{66.7}$ (R = La–Nd) and $La_{22.2}M_{11.1}Ge_{66.7}$ (M = Cu, Ag) nominal compositions were prepared by direct synthesis in resistance furnace. Stoichiometric amounts of components were placed in an arc–sealed Ta crucible in order to prevent oxidation; the constituents of the Li–containing alloys and the $Eu_{22.2}Pd_{11.1}Ge_{66.7}$ sample were weighed and sealed in the crucible inside an inert atmosphere glove box. The

---

[1]Part of this section was originally published in [35].

Ta crucible was then closed in an evacuated quartz phial to prevent oxidation at high temperature, and finally placed in a resistance furnace, where the following thermal cycle was applied:

(I)   25 °C → (10 °C/min) → 950 °C (1 h) → (−0.2 °C/min) → 350 °C furnace switched off

A continuous rotation, at a speed of 100 rpm, was applied to the phial during the thermal cycle. These synthetic conditions were chosen with the aim to obtain samples containing crystals of good quality and size, suitable for further structural studies. The synthesized alloys are very brittle and mostly stable in air, except for the $Eu_{22.2}Pd_{11.1}Ge_{66.7}$ and the Li–containing ones. No tantalum contaminations of the samples were detected.

After direct synthesis in the resistance furnace, some pieces of the $La_{22.2}Pd_{11.1}Ge_{66.7}$ sample were annealed at different temperatures. Further attempts to synthesize the La_2PdGe_6 compound were done, by arc melting of components, followed by annealing treatments (see paragraph 4.3.1) planned after thermal effects detected by DTA.

Flux synthesis was also performed for samples containing La, Pr and Yb, using In as metal solvent. Stoichiometric amounts of R, Pd and Ge, giving the nominal composition $R_{21}Pd_7Ge_{72}$, were put in an arc–sealed Ta crucible with a 1:45 molar excess of indium, to obtain a total mass of about 3 g. This composition was chosen in order to avoid unwanted side reactions between In and Pd. Then, the Ta crucible, closed in an evacuated quartz phial, was placed in a resistance furnace, and the following thermal cycles were applied:

(II)   25 °C → (2 °C/min) → 1000 °C (5 h) → (−1.0 °C/min) → 850 °C (48 h) → (−0.3 °C/min) → 25 °C

(III) 25 °C → (10 °C/min) → 750 °C (24 h) → (−0.5 °C/min) → 25 °C

For the La–containing representative both thermal cycles were tested, instead only the cycle III was applied to the Pr and Yb–containing samples.

The quartz phial was kept under continuous rotation during the thermal cycle aiming to favour a better dissolution of the constituting elements inside the flux. In all cases, a vertical cut of the obtained ingots revealed the presence of large shining crystals, visible to the naked eye, randomly distributed within the flux solidified matrix.

In order to perform the metallographic analysis, samples were prepared as described in paragraph 2.2.

After SEM–EDXS analysis, crystals of R_2PdGe_6 compounds ($R =$ La, Pr, Yb) obtained by flux synthesis were extracted from the flux by immersion and sonication of the ingot in glacial acetic acid [$CH_3COOH_{(l)}$] for about 24 h, allowing indium selective oxidation. The obtained crystalline product was rinsed with water and dried with acetone. After that, another SEM–EDXS analysis was performed in order to check the goodness of the separation procedure and the quality/composition of the isolated crystals.

Microstructure examination as well as qualitative and quantitative analyses were performed by SEM–EDXS using cobalt standard for calibration. It is important to highlight the problem of dealing with Li samples since it is not detectable by the EDXS. Thus, for instance, the wanted R_2LiGe_6 phase was detected as a binary "RGe_3" (composition in *at*% equal to $R_{25.0}Ge_{75.0}$). In these cases, the structural information from diffraction experiments on both powder and single crystals are of crucial importance.

4.2   Computational Details[2]

DFT total energy calculations were performed for R_2PdGe_6 ($R = $ Y, La) in the three structural modifications $oS18$, $oS72$ and $mS36$, and for La_2MGe_6 ($M = $ Li, Pt, Cu, Ag, Au) in the two orthorhombic modifications, by means of the plane wave pseudopotential code QUANTUM–ESPRESSO [37]. PBE functional [38] for the exchange and correlation energy were used. Ultrasoft pseudopotentials [39], available in the "GBRV" open–source library [40], were employed for M, La and Y, whereas for Ge a norm–conserving pseudopotential, including the 4 s and 4p valence orbitals, was used. The semicore states 4p for Pd, 1 s for Li, 5p for Pt, 3 s and 3p for Cu, 4 s and 4p for Ag and for Y and 5 s and 5p for La, were treated as valence electrons. The Brillouin zone was sampled within uniform grids generated with different k–points for the three polymorphs: $12 \times 12 \times 2$ for the $oS18$ modification, $6 \times 6 \times 2$ for the $oS72$ and $6 \times 6 \times 4$ for the $mS36$. The plane–wave and density cut–off were set to 45 Ry and 450 Ry, respectively. Orbital occupancies at the Fermi level were treated with a Gaussian smearing of 0.01 Ry.

In order to perform the position–space chemical bonding analysis, the electronic structure calculations for the crystalline compounds La_2MGe_6 ($M = $ Li, Mg, Al, Zn, Cu, Ag, Pd) and Y_2PdGe_6 have been effectuated at the DFT/PBE level of theory (program FHIaims [41, 42]) using the literature reported (for the Al, Mg and Zn analogues) and the here experimentally determined (for the others) structure parameters. Again, La and Y were chosen in order to avoid partially filled 4f states. For the Brillouin zone sampling a (4 4 2) and a (4 4 4) k–point mesh were used for the $oS72$ and the $mS36$ structures respectively. For all species the preconstructed default "tight" basis set were chosen and scalar relativistic effects for all electrons were taken into account within the ZORA approach. The orbitals occupancies at the Fermi level were treated with a Gaussian smearing of 0.01 eV.

The additional chemical bonding investigations performed on $LaPdGe_3$ and $CaGe$, using literature crystal data [43, 44], were performed with the same software and settings as for the R_2MGe_6 phases. The Brillouin zone was sampled with a (8 8 4) and a (6 2 6) k–point mesh for $LaPdGe_3$ and $CaGe$ respectively.

A position–space chemical bonding analysis was also performed for La_2MgGe_6 and La_2ZnGe_6, by means of the FPLO [45] program package. The influence of

---

[2]Part of this section was originally published in [35,36].

both Perdew–Wang (PW92) [46] and PBE [38] exchange–correlation potentials were tested for the LSDA and GGA calculations, respectively. The energy convergence with respect to the number of k–points was also tested and reached using (12 12 6) k–points. As already mentioned, FPLO doesn't give access to the wave function, precluding, for instance, the possibility to perform electron density refinement and the evaluation of fields like the ELI–D Laplacian. In addition, the Ge lone–pair basins overpopulation, with respect to bonding basins, (see paragraph 4.6) was always higher than for FHI–aims results, also after the PSC0 correction. That is why, the complete chemical bonding analysis was performed on the basis of FHI–aims wave functions. Anyway, results for La_2MgGe_6 and La_2ZnGe_6, without and after the PSC0, obtained through FPLO (PBE/12–12–6 k–mesh), are listed in Tables A4.1–A4.3.

4.3 On the Formation of R_2MGe_6 Germanides Along the R Series

4.3.1 The R_2PdGe_6 (R = Sc, Y, La–Nd, Sm–Lu) Samples[3]

Results of SEM/EDXS and XRPD characterization confirm the presence of the R_2PdGe_6 phase in the most part of samples prepared by direct synthesis (see Table 4.2).

According to SEM micrographs, under the applied experimental conditions for direct synthesis, the R_2PdGe_6 is generally the highest yield phase. As it is common for alloys prepared by slow cooling from liquid, many secondary phases were detected, among which $R(Pd,Ge)_{2-x}$, $R_2Pd_3Ge_5$, $RPdGe_2$ and Ge are the most common and others which still need to be investigated. It should be noted that during the study on R_2PdGe_6 series, the new $Nd_2Pd_3Ge_5$ compound was found and its crystal structure solved: these results are deeply discussed in Chap. 5.

Representative microstructures and X–ray powder pattern for a light (Nd) and heavy (Yb) rare earth representatives are reported in Fig. 4.4.

In samples with R = Sc and Eu no traces of the R_2PdGe_6 phase were detected: in both cases three–phase alloys were obtained, in which an already known ternary compound (ScPdGe or EuPdGe$_3$) coexists with Ge and the binary RGe_{2-x} phase dissolving, in particular with Eu, a small amount of Pd.

In the $La_{22.2}Pd_{11.1}Ge_{66.7}$ sample prepared by direct synthesis applying cycle I (sample code "La" in Table 4.2), only a small amount of La_2PdGe_6 was found, in the form of a thin border around big $LaGe_{2-x}$ crystals (see Fig. 4.5a). From such a sample, it was not possible to isolate suitable single crystals. In the literature, this phase is reported to crystallize in the $oS18$–Ce_2CuGe_6 [2] or in the $oS72$– $Ce_2(Ga_{0.1}Ge_{0.9})_7$ model [6], the latter with no complete structural data available. For this reason,

---

[3]Part of this section was originally published in [35].

Table 4.2 Structure data and elemental atomic per cent composition for $R_{22.2}Pd_{11.1}Ge_{66.7}$ samples prepared by direct synthesis in resistance furnace applying cycle I and for the $R_{21}Pd_7Ge_{72}$ samples prepared by flux synthesis using cycle III (indicated with f in the sample code)

Sample code (R)	Phases	Composition by (EDXS) [at %]			Pearson symbol prototype	Lattice parameters [Å]		
		R	Pd	Ge		a	b	c
Sc	ScGe$_2$	34.2	–	65.8	$oS12$–ZrSi$_2$	3.8787(8)	14.876(2)	3.7952(4)
	ScPdGe	34.7	30.5	34.8	$hP9$–ZrNiAl	6.704(1)	6.704(1)	3.958(2)
	Ge	–	–	100	$cF8$–C	5.654		
Y	**Y$_2$PdGe$_6$** °	**21.0**	**10.8**	**68.2**	$oS72$–Ce$_2$(Ga$_{0.1}$Ge$_{0.9}$)$_7$	**8.1703(5)**	**8.0451(5)**	**21.558(1)**
	YPdGe$_2$ *	23.5	24.8	51.7	$oI32$–YIrGe$_2$			
	New phase I	18.9	37.8	43.3				
	New phase II	31.4	20.6	48.0				
	New phase III	29.1	7.8	63.1				
La	**La$_2$PdGe$_6$**	**22.4**	**9.9**	**67.7**	$oS72$–Ce$_2$(Ga$_{0.1}$Ge$_{0.9}$)$_7$	**8.433(3)**	**8.208(7)**	**22.194(9)**
	La(Pd,Ge)$_{2-x}$ *	35.2	1.5	63.3	$oI12$–GdSi$_{2-x}$			
	LaPdGe$_3$	20	19.5	60.5	$tI10$–BaNiSi$_3$	4.5077(5)	4.5077(5)	9.849(1)
	Ge	–	–	100	$cF8$–C	5.654		
La, f	**La$_2$PdGe$_6$** °	**22.4**	**10.5**	**67.1**	$mS36$–La$_2$AlGe$_6$	**8.2163(8)**	**8.4161(9)** $\beta=100.477(1)$	**11.294(1)**
	La(Pd,Ge)$_{2-x}$ #	34.5	3.3	62.2	$cF8$–C			
	Ge#	–	–	100				
Ce	**Ce$_2$PdGe$_6$** °	**22.5**	**10.8**	**66.7**	$oS72$–Ce$_2$(Ga$_{0.1}$Ge$_{0.9}$)$_7$	**8.3548(4)**	**8.1774(4)**	**22.0272(9)**
	Ce$_2$Pd$_3$Ge$_5$ *	20.2	29.2	50.6	$oI40$–U$_2$Co$_3$Si$_5$			
	New phase I	24.1	20.3	55.6				
	Ge	–	–	100	$cF8$–C	5.654		

(continued)

Table 4.2 (continued)

Sample code (R)	Phases	Composition by (EDXS) [at %]			Pearson symbol prototype	Lattice parameters [Å]		
		R	Pd	Ge		a	b	c
Pr	**Pr$_2$PdGe$_6$**°	**20.2**	**11.4**	**68.4**	**oS72–Ce$_2$(Ga$_{0.1}$Ge$_{0.9}$)$_7$**	**8.3129(5)**	**8.1540(5)**	**21.996(1)**
	Pr$_2$Pd$_3$Ge$_5$*	19.1	29.6	51.3	oI40–U$_2$Co$_3$Si$_5$			
	New phase I	22.4	20.0	57.6				
	New phase II	30.6	9.9	59.5				
	Ge	–	–	100	cF8–C	5.654		
Pr, f	**Pr$_2$PdGe$_6$** °	**20.6**	**11.3**	**68.1**	**mS36–La$_2$AlGe$_6$**	**8.157(2)**	**8.328(2)** β=100.391(3)	**11.195(2)**
	Pr(Pd,Ge)$_{2-x}$#	33.5	1.0	65.5				
	Ge#	–	–	100				
Nd	**Nd$_2$PdGe$_6$**°	**21.8**	**11.2**	**67.0**	**oS72–Ce$_2$(Ga$_{0.1}$Ge$_{0.9}$)$_7$**	**8.2831(4)**	**8.1323(4)**	**21.915(1)**
	Nd$_2$Pd$_3$Ge$_5$ °	19.2	29.7	51.1	oI40–U$_2$Co$_3$Si$_5$	10.1419(8)	6.1317(5)	12.056(1)
	New phase I	31.9	11.9	56.2				
	NdPdGe$_2$#	24.2	24.8	51.0				
	New phase II	27.4	12.2	60.4				
	Ge	–	–	100	cF8–C	5.654		
Sm	**Sm$_2$PdGe$_6$**	**21.5**	**11.1**	**67.4**	**oS72–Ce$_2$(Ga$_{0.1}$Ge$_{0.9}$)$_7$**	**8.209(1)**	**8.072(1)**	**21.630(5)**
	Sm$_2$Pd$_3$Ge$_5$#	19.3	29.9	50.8				
	New phase I	31.8	5.7	62.5				
	Ge	–	–	100	cF8–C	5.654		

(continued)

Table 4.2 (continued)

Sample code (R)	Phases	Composition by (EDXS) [at %]			Pearson symbol prototype	Lattice parameters [Å]		
		R	Pd	Ge		a	b	c
Eu	EuPdGe₃	19.2	20.0	60.8	$tI10$– BaNiSi₃	4.4459(7)	4.4459(7)	10.153(2)
	EuGe₂₋ₓ*	32.1	–	67.9	$hP3$– AlB₂			
	Ge	–	–	100	$cF8$–C	5.654		
Gd	**Gd₂PdGe₆**	**22.2**	**11.4**	**66.4**	$oS72$–Ce₂(Ga₀.₁Ge₀.₉)₇	**8.188(1)**	**8.065(1)**	**21.575(6)**
	New phase I	33.1	4.5	62.4				
	Gd(Pd,Ge)₂₋ₓ#	37.2	0.4	62.4				
	New phase I	29.8	8.7	61.5				
	New phase II	20.6	29.4	50.0				
Tb	**Tb₂PdGe₆**	**22.2**	**11.1**	**66.7**	$oS72$–Ce₂(Ga₀.₁Ge₀.₉)₇	**8.1533(9)**	**8.0045(8)**	**21.547(4)**
	New phase I	29.7	8.3	62.0				
	New phase II	19.7	37.9	42.4				
	TbPdGe₂#	24.7	24.3	51.0				
Dy	**Dy₂PdGe₆**	**22.7**	**10.9**	**66.4**	$oS72$–Ce₂(Ga₀.₁Ge₀.₉)₇	**8.140(1)**	**8.016(3)**	**21.484(5)**
	DyPdGe₂#	25.1	25.3	49.6				
	New phase I	31.4	8.1	60.5				
	New phase II	20.4	38.4	41.2				
	New phase III	32.8	21.1	46.1				
Ho	**Ho₂PdGe₆**	**22.3**	**11.3**	**66.4**	$oS72$–Ce₂(Ga₀.₁Ge₀.₉)	**8.1251(7)**	**8.0107(9)**	**21.431(2)**
	New phase I	26.8	30.1	43.1				
	Ho(Pd,Ge)₂₋ₓ#	30.4	7.6	62.0				

(continued)

Table 4.2 (continued)

Sample code (R)	Phases	Composition by (EDXS) [at %]			Pearson symbol prototype	Lattice parameters [Å]		
		R	Pd	Ge		a	b	c
Er	**Er$_2$PdGe$_6$** °	**22.7**	**11.0**	**66.3**	$oS72$–Ce$_2$(Ga$_{0.1}$Ge$_{0.9}$)$_7$	**8.1122(6)**	**8.0068(6)**	**21.399(2)**
	Er(Pd,Ge)$_{2-x}$#	31.8	6.6	61.6				
	New phase I	27.4	30.8	41.8				
	New phase II	20.6	38.6	40.8				
	New phase III	19.8	22.4	57.8				
	New phase IV	39.2	0.4	60.4				
Tm	**Tm$_2$PdGe$_6$**	**23.0**	**11.1**	**65.9**	$oS72$–Ce$_2$(Ga$_{0.1}$Ge$_{0.9}$)$_7$	**8.072(5)**	**7.986(5)**	**21.279(9)**
	Tm(Pd,Ge)$_{2-x}$#	32.2	7.2	60.6				
	New phase II	28.5	30.8	40.7				
Yb	**Yb$_2$PdGe$_6$**	**23.1**	**10.7**	**66.2**	$oS72$–Ce$_2$(Ga$_{0.1}$Ge$_{0.9}$)$_7$	**8.138(2)**	**7.975(2)**	**21.836(9)**
	Yb(Pd,Ge)$_{2-x}$*	35.0	16.6	48.4	$hP3$– AlB$_2$			
Yb,f	**Yb$_2$PdGe$_6$** °	**23.2**	**10.9**	**65.9**	$oS72$–Ce$_2$(Ga$_{0.1}$Ge$_{0.9}$)$_7$	**8.1428(3)**	**7.9807(3)**	**21.8331(9)**
	YbGe$_{2-x}$#	38.4	–	61.6				
Lu	**Lu$_2$PdGe$_6$** °	**23.8**	**10.6**	**65.6**	$oS72$–Ce$_2$(Ga$_{0.1}$Ge$_{0.9}$)$_7$	**8.0725(8)**	**7.9791(8)**	**21.317(2)**
	New phase I	30.9	22.7	46.4				
	Lu(Pd,Ge)$_{2-x}$#	33.6	4.9	61.5				
	Ge	–	–	100.0	$cF8$–C	5.654		

o Phases the structure and lattice parameters of which were determined by means of X-ray single crystal analysis

* Phases identified by XRPD analysis

Lattice parameters were not refined for these phases due to their low amount and/or strong peak overlapping

Phases identified only by EDXS analysis

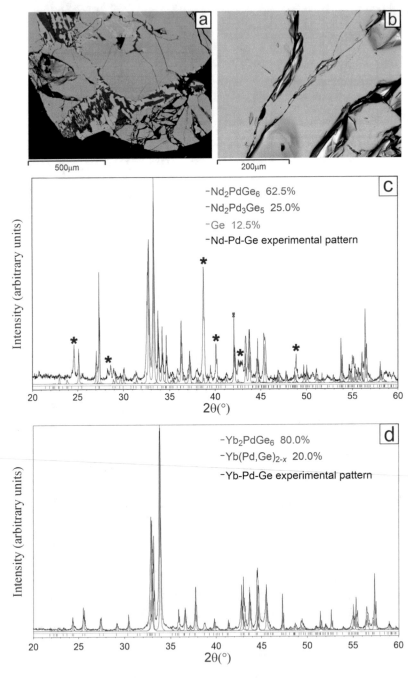

Fig. 4.4 BSE micrographs and X–ray powder patterns for **a**, **c** the Nd and **b**, **d** the Yb representatives. **a** Nd_2PdGe_6 (grey phase); $Nd_2Pd_3Ge_5$ (bright phase); Ge (dark phase); **b** Yb_2PdGe_6 (grey phase). No other phases are visible in this area. The symbol * in **c** indicates non indexed peaks

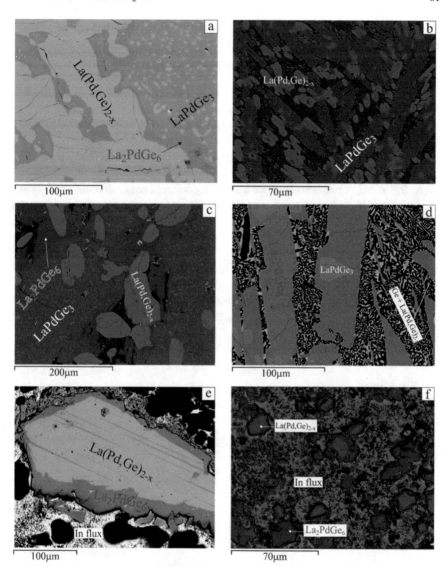

Fig. 4.5 Micrographs (SEM–BSE mode) of representative samples of $La_{22.2}Pd_{11.1}Ge_{66.7}$ nominal composition (**a–d**) and of $La_{21}Pd_7Ge_{72}$ nominal composition (**e, f**) obtained under the following experimental conditions: **a** synthesis in resistance furnace, cycle I; **b** arc melting; **c–d** arc melting followed by one DTA cycle; **e** synthesis in resistance furnace with In flux, cycle II; **f** synthesis in resistance furnace with In flux, cycle III. From [35]

further attempts were done to synthesize a La_2PdGe_6 sample convenient for structural resolution, which are briefly accounted for in the following.

As–cast samples with nominal composition $La_{22.2}Pd_{11.1}Ge_{66.7}$ did not show any trace of La_2PdGe_6, independently from the applied melting method (arc or induction furnace). These samples are characterized by a clear microstructure, where crystals of $La(Pd,Ge)_{2-x}$ and $LaPdGe_3$ coexist with an eutectic structure containing Ge (see Fig. 4.5b). Annealing at 700 °C of an as–cast sample for 2 weeks, performed on the basis of literature data [6], did not succeed in obtaining the desired compound; what is more, the same annealing treatment carried out on the sample (La) prepared by cycle I caused the thin La_2PdGe_6 border to disappear.

At this point, it became clear that La_2PdGe_6 behaves somewhat differently from the other rare earth homologues, and DTA measurements were performed in order to gain insights on its temperature formation.

The DTA curve obtained with heating rate of 5 °C/min ($T_{max} = 1100$ °C), is shown in Fig. 4.6.

Four thermal effects were detected at the following temperatures (onset): 800 °C, 886 °C, 927 °C and 1027 °C. The same thermal effects are recorded on cooling. The peak at 800 °C is in good agreement with the temperature of the binary eutectic equilibrium $L \rightarrow (Ge) + LaGe_{2-x}$ reported at T $= 810$ °C [47]. However, the other peaks are not easily interpretable. After this thermal cycle the presence of a small amount of La_2PdGe_6 was indeed detected by SEM–EDXS analysis in the form of a

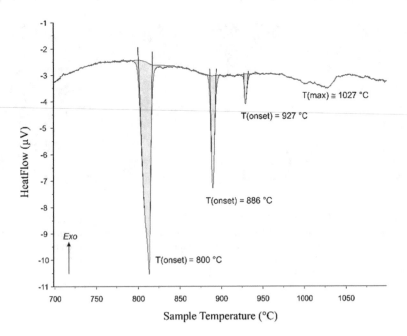

Fig. 4.6 Differential Thermal Analysis curve (heating regime; 5 °C/min) for an arc melted sample of nominal composition $La_{22.2}Pd_{11.1}Ge_{66.7}$

border between $La(Pd,Ge)_{2-x}$ and $LaPdGe_3$ (see Fig. 4.5c). In a restricted region of the sample after DTA, bigger crystals of La_2PdGe_6 are present enclosing a brighter core of $LaPdGe_3$ (see Fig. 4.5d). Aiming to obtain a higher yield of the desired compound, different annealing treatments (lasting one month each) were performed at 830 °C, 890 °C and 1000 °C: temperatures were chosen slightly below the recorded thermal effects. Unfortunately, no traces of 2:1:6 were found after all these treatments. From the gathered results, it was concluded that La_2PdGe_6 is probably a metastable phase which is likely to form only in small amount during relatively slow cooling.

It is known that alternative synthetic routes may help in producing metastable compounds; among them, the flux method was targeted, taking into account both literature data [48] and our previous results on related $R_2Pd_3Ge_5$ compounds [49] (see Chap. 5). As a metal solvent In was chosen considering its ability to dissolve Ge, rare earth and transition metals, without forming In–Ge binary compounds, allowing the formation of ternary phases [48, 50]. Thus, several attempts were done varying both the nominal composition and the thermal cycle. Good results were obtained starting from the nominal composition $La_{21}Pd_7Ge_{72}$: the 2:1:6 phase was obtained applying both cycle II and cycle III (see Fig. 4.5e and f). After cycle II, however, La_2PdGe_6 is almost always found as a grey border of big $La(Pd,Ge)_{2-x}$ crystals and only a few single phase crystals were detected. On the contrary, after cycle III, characterized by a lower maximum temperature (750 °C instead of 1000 °C), many 2:1:6 single crystals were obtained, allowing further crystal structure determination after their separation from the In flux (sample indicated by code "La, f" in Table 4.2). All the detailed information about the synthetic attempts to prepare a sufficient yield of La_2PdGe_6 for structural study, can be found in the Appendix section, Table A4.4.

Unexpectedly, the flux synthesized La_2PdGe_6 turned out to be monoclinic, differently from all the other series representatives prepared by direct synthesis (see paragraph 4.3): for this reason, the flux method (cycle III) was tested also on $Pr_{21}Pd_7Ge_{72}$

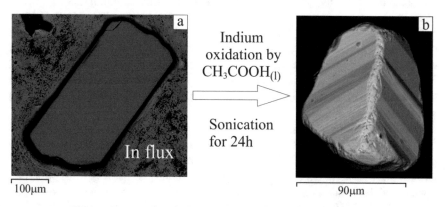

Indium oxidation by $CH_3COOH_{(l)}$

Sonication for 24h

In flux

100µm

90µm

Fig. 4.7 SEM–BSE micrographs for Yb_2PdGe_6 single crystal before (**a**) and after (**b**) the selective In oxidation. From [35]

and $Yb_{21}Pd_7Ge_{72}$ samples. In both cases R_2PdGe_6 phases were detected and analysed by X–ray diffraction (see Table 4.2); particularly good quality big crystals were obtained in the case of R = Yb (see Fig. 4.7).

4.3.2   The R_2LiGe_6 (R = La–Nd) and La_2MGe_6 (M = Cu, Ag) Samples

As it was already mentioned, the compounds with general formula R_2LiGe_6 (R = La–Pr) were already studied and reported with another $oS18$ structure, unique for the Li analogues, with the Pr_2LiGe_6 representative as prototype. The existence and crystal structure of this series was investigated, including also the $Nd_{22.2}Li_{11.1}Ge_{66.7}$ alloy. Results of SEM/EDXS and XRPD characterization are listed in Table 4.3.

Phases with EDXS composition close to the expected $R_{25.0}Ge_{75.0}$ were detected in all the synthesized sample. Anyway, under the applied conditions, good quality X–ray powder pattern were obtained only with La, where the wanted La_2LiGe_6 turned out to be the highest yield phase (see Fig. 4.8a and c). After X–ray single crystal diffraction, the $mS36$–La_2AlGe_6 structure was assigned to it, allowing to revise the previously reported data. The lowest yield phase, detected only by SEM/EDXS, with $La_{28.9}Ge_{78.1}$ composition was interpreted as La_4LiGe_{10-x} on the basis of other results obtained during the doctorate work (see Chap. 7).

With R = Ce, Pr the most intense Bragg peaks were indexed with the same $mS36$ modification, confirming the existence of Ce_2LiGe_6 and Pr_2LiGe_6. Anyway, the correct structure type is not deducible from such samples and was assigned according to results obtained with the La analogue. In fact, no structure transitions were reported along the series with other M metals within the 2:1:6 family of intermetallics. Nevertheless, single crystal diffraction experiments would definitely confirm these hypotheses. The situation is somewhat different with Nd. In this case, the highest yield phase is Nd_4LiGe_{10-x} which was also successfully indexed (Fig. 4.8d). Although in the powder pattern none of the remaining peak can be assign to the 2:1:6 compound, a $Nd_{23.8}Ge_{76.2}$ composition phase was detected by means of EDXS in one small piece of the analysed sample (Fig. 4.8b). This phase could be the new Nd_2LiGe_6. To confirm it new syntheses, maybe with Ge poorer nominal composition in order to avoid the formation of Nd_4LiGe_{10-x}, should be planned and will be the object of further investigations.

These results represent one more step on the road to clarify the allowed structural type for the 2:1:6 stoichiometry phases. Further study, on the basis of symmetry principle and theoretical DFT results, were performed and described in the following. After that, in order to experimentally check the validity of the suggested structural revision of the whole family of the R_2MGe_6 phases, two intermetallics previously reported [2, 51, 52] to crystalize with the $oS18$–Ce_2CuGe_6 structure were synthesized (cycle I) and characterized: La_2CuGe_6 and La_2AgGe_6. X–ray powder diffraction

Table 4.3 Structure data and elemental atomic per cent composition for $R_{22.2}Li_{11.1}Ge_{66.7}$ samples prepared by direct synthesis in resistance furnace applying cycle I

Sample code (R)	Phases	Composition by (EDXS) [at %]		Pearson symbol prototype	Lattice parameters [Å]		
		R	Ge		a	b	c
La	**La$_2$LiGe$_6$°**	**25.1**	**74.9**	**mS36-La$_2$AlGe$_6$**	**8.3597(3)**	**8.8595(3) β=100.113(1)**	**10.8539(4)**
	La$_4$LiGe$_{10-x}$#	28.9	71.1				
	Ge	–	100		5.654		
Ce	**Ce$_2$LiGe$_6$**	**25.6**	**74.4**	**mS36-La$_2$AlGe$_6$**	**8.19(1)**	**8.745(9) β=101.7(1)**	**10.83(1)**
	Ce$_4$LiGe$_{10-x}$#	30.2	69.8				
	CeGe#	50.8	49.2				
	Ge	–	100	cF8-C	5.654		
Pr	**Pr$_2$LiGe$_6$**	**23.8**	**76.2**	**mS36-La$_2$AlGe$_6$**	**8.24(1)**	**8.64(1) β=101.7(1)**	**10.661(1)**
	Pr$_4$LiGe$_{10-x}$#	27.4	72.6				
	Ge	–	100	cF8-C	5.654		
Nd	**Nd$_2$LiGe$_6$#**	**23.8**	**76.2**				
	Nd$_4$LiGe$_{10-x}$	28.8	71.2	mS60-y-La$_4$MgGe$_{10-x}$	8.596(6)	8.315(2) β=97.08(4)	17.559(7)
	~ NdGe$_2$	34.6	65.4				

o Phases whose structure and lattice parameters were determined by means of X–ray single crystal analysis

* Phases identified by XRPD analysis

Lattice parameters were not refined for these phases due to their low amount and/or strong peak overlapping

Phases identified only by EDXS analysis

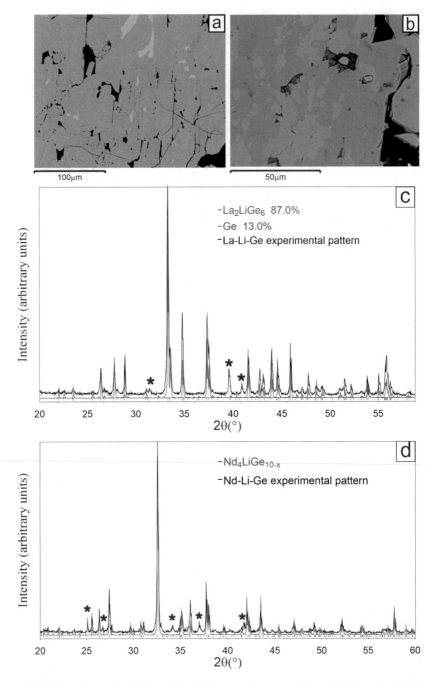

Fig. 4.8 SEM micrographs (BSE mode) and X–ray powder patterns for **a, c**) the La and **b, d**) the Nd representatives. **a** La_2LiGe_6 (grey phase); La_4LiGe_{10-x} (bright phase); **b** Nd_2LiGe_6 (grey phase); Nd_4LiGe_{10-x} (bright phase). The symbol * indicates non indexed peaks

results (Fig. 4.9), together with unit cell parameters, as obtained on the basis of least square refinement from powder data, are listed in Table 4.4.

Analogously to the Pd phases, also the Cu and Ag containing germanides crystallize with the $oS72$ structure, and not again with the $oS18$. Detailed on their structure solutions and chemical bonding are reported in the following paragraphs.

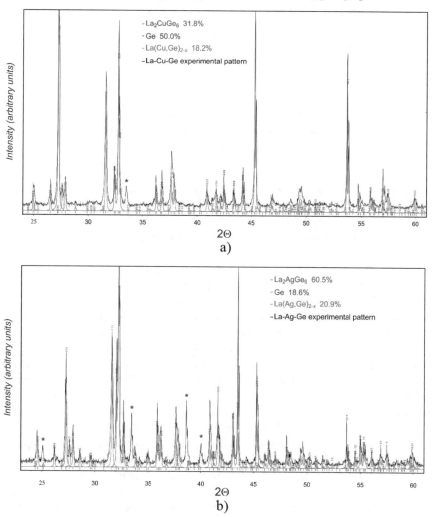

Fig. 4.9 X–ray powder patterns for **a** La_2CuGe_6 and **b** La_2AgGe_6. The symbol * indicates non indexed peaks

Table 4.4 Structure data for $La_{22.2}M_{11.1}Ge_{66.7}$ ($M =$ Cu, Ag) samples prepared by direct synthesis in resistance furnace applying cycle I

Sample code (M)	Phases	Pearson symbol prototype	Lattice parameters [Å]		
			a	b	c
Cu	**La_2CuGe_6**	$oS72$–$Ce_2(Ga_{0.1}Ge_{0.9})_7$	**8.505(1)**	**8.186(1)**	**21.789(5)**
	$La(Cu,Ge)_{2-x}$	$oI12$–$GdSi_{2-x}$	4.320(3)	4.406(1)	14.206(5)
	Ge	$cF8$–C	5.654		
Ag	**La_2AgGe_6**	$oS72$–$Ce_2(Ga_{0.1}Ge_{0.9})_7$	**8.676(4)**	**8.324(5)**	**21.896(5)**
	$La(Ag,Ge)_{2-x}$	$tI12$–$GdSi_{2-x}$	4.318(1)	4.4153(6)	14.190(2)
	Ge	$cF8$–C	5.654		

4.4   Crystal Structure of R_2PdGe_6 ($R =$ Y, La (Flux), Ce, Pr, Pr (Flux), Nd, Er, Yb(Flux), Lu) and La_2MGe_6 ($M =$ Li, Cu, Ag)[4]

Single crystals of R_2PdGe_6, both prepared by direct synthesis ($R =$ Y, Ce, Pr, Nd, Er, Lu) and by flux synthesis ($R =$ La, Pr, Yb), and of La_2MGe_6 ($M =$ Li, Cu, Ag) compounds were selected with the aid of a light optical microscope operated in the dark field mode and analyzed by means of X–ray single crystal diffractometer as described in paragraph 2.3.2.

For single crystals of R_2PdGe_6 ($R =$ Y, Ce, Pr, Nd, Er, Yb(f), Lu) and La_2MGe_6 ($M =$ Cu, Ag) compounds the cell indexation was straightforward giving an orthorhombic C–centered cell (only $h + k = 2n$ reflections were observed). The analysis of systematic extinctions suggested as possible space groups $Cc2e$ (№ 41) and $Cmce$ (№ 64). An almost complete structural model was obtained in the $Cmce$ space group in a few iteration cycles by applying the charge–flipping algorithm implemented in $JANA2006$ [53]. In this model the rare earth atoms are situated in a 16 g general site, the palladium/copper/silver species occupy the $8f$ site, while all the other positions are assigned to the lighter germanium atoms. The obtained models have $oS72$ Pearson symbol and correspond to the R_2PdGe_6 stoichiometry, satisfactorily matching with EDXS microprobe analysis data.

The further structure refinements were carried out by full–matrix least–squares methods on $|F^2|$ using the SHELX programs [54]. The site occupancy factors of all species were checked for deficiency, in separate cycles of refinement, obtaining values very close to unity. At this point neither deficiency nor statistical mixture were considered and stoichiometric R_2MGe_6 models were further anisotropically refined giving acceptable residuals and flat difference Fourier maps. Results indicate that the R_2PdGe_6 ($R =$ Y, Ce, Pr, Er, Lu), La_2CuGe_6 and La_2AgGe_6 compounds prepared by direct synthesis are isopointal with the $Ce_2(Ga_{0.1}Ge_{0.9})_7$ prototype. The same is true also for the Yb_2PdGe_6 containing compound prepared by flux synthesis (the image of this crystal is reported in Fig. 4.7).

---

[4]Part of this section was originally published in [35].

Selected crystallographic data and structure refinement parameters for these crystals are listed in Tables 4.5 and 4.6.

The atomic positions of the R_2PdGe$_6$, La$_2$CuGe$_6$ and La$_2$AgGe$_6$ compounds are listed in Tables 4.7 and 4.8 together with the equivalent isotropic displacement parameters. Positions were standardized with the Structure Tidy program [55]. Interatomic distances are available in the Appendix (Table A4.5).

The La$_2$PdGe$_6$ crystal selected for X–ray analysis, taken from the flux synthesized sample, is an example of non–merohedral twin composed of two domains of comparable dimensions. Normally, twins of such type give problems on the preliminary stages of cell indexing (leading to unexpectedly high values of unit cell parameters) and space group determination (showing inconsistency with any known space group systematic absences) [56, 57]. Instead, in our case the unit cell indexing was straightforward and the following possible space groups were suggested for the C–centered monoclinic cell: $C2$ (№ 5), Cm (№ 8) and $C2/m$ (№ 12). The lowest combined figure of merit was associated with the only centrosymmetric $C2/m$ space group. These data strongly hint that the studied crystal is isostructural with the monoclinic $mS36$–La$_2$AlGe$_6$ prototype. A preliminary structural model, obtained by direct methods as implemented in *WinGx* [58], contained 1 La, 5 Ge and 1 Pd, giving the correct La$_2$PdGe$_6$ composition. Nevertheless, the isotropic thermal parameters values were not coherent for the different sort of atoms and three additional intense peaks close to some of the Ge–positions were present in the difference Fourier maps at distances ~ 0.05 nm. The latter three sites have no physical sense if completely occupied; therefore, the sum of occupations for each pair of very close Ge–sites was restrained to be unity in further cycles of least squares refinement. Anyway, the refinement sticks at R1/wR2 of 0.05/0.17 with a noisy Fourier map, giving unreasonable thermal parameters when anisotropically refined. Even less chemically sound structural models were obtained testing the possible non–centrosymmetric $C2$ and Cm space groups. At this point, a more careful analysis of diffraction spots in reciprocal space was performed with RLATT and CELL_NOW [59]. In fact, a regular spatial distribution of extra–peaks was revealed with respect to those associated with $mS36$–like monoclinic cell. All of them could be satisfactory indexed with a twice as big monoclinic base centered unit cell with $a \sim 8.2$ Å, $b \sim 8.4$ Å, $c \sim 22.6$ Å, and $\beta \sim 100.5°$. The structural model deduced in $C2/m$ space group by the charge–flipping algorithm [53] became less disordered, since it contains only one partially occupied Ge position capping the distorted corrugated Ge fragment [14]. Even if Ge–rich compounds are frequently off–stoichiometric or disordered [1], residuals remain unsatisfactory (0.09/0.14) with senseless anisotropic thermal parameters. For this reason, more attention was dedicated to the indexing procedure. According to a detailed output of CELL_NOW, one of the possible interpretations of the observed diffraction peaks distribution is considering them originating from different domains of a non–merohedrically twinned crystal. In this case, the metric of the single domain remains

Table 4.5 Crystallographic data for R_2PdGe_6 (R = Y, Ce, Pr, Nd, Er and Lu) single crystals taken from samples prepared by direct synthesis and experimental details of the structural determination

Empirical formula	Y_2PdGe_6	Ce_2PdGe_6	Pr_2PdGe_6	Nd_2PdGe_6	Er_2PdGe_6	Lu_2PdGe_6
EDXS data	$Y_{21.0}Pd_{10.8}Ge_{68.2}$	$Ce_{22.2}Pd_{10.8}Ge_{67.0}$	$Pr_{20.2}Pd_{11.4}Ge_{68.4}$	$Nd_{21.8}Pd_{11.2}Ge_{67.0}$	$Er_{22.7}Pd_{11.0}Ge_{66.3}$	$Lu_{23.8}Pd_{10.6}Ge_{65.6}$
M_w, [g/mol]	719.76	822.18	823.76	830.42	876.46	891.88
a [Å]	8.1703(5)	8.3548(4)	8.3129(5)	8.2831(4)	8.1122(6)	8.0725(8)
b [Å]	8.0451(5)	8.1774(4)	8.1540(5)	8.1323(4)	8.0068(6)	7.9791(8)
c [Å]	21.558(1)	22.0272(9)	21.996(1)	21.915(1)	21.399(2)	21.317(2)
V [Å³]	1417.1(2)	1504.9(1)	1491.0(2)	1476.2(1)	1389.9(2)	1373.0(2)
Calc. density [g/cm³]	6.748	7.258	7.340	7.473	8.377	8.629
abs coeff (μ), mm⁻¹	43.6	37.7	38.9	40.1	51.8	56.8
Unique reflections	1028	1308	1296	1307	1206	1225
Reflections I > 2σ(I)	884	1072	870	848	1020	874
R_{sigma}	0.0233	0.0084	0.0270	0.0263	0.0133	0.0284
Data/parameters	884/50	1072/50	870/50	848/50	1020/50	874/50
GOF on F² (S)	1.166	1.543	1.059	1.002	1.326	0.912
R indices [I > 2σ(I)]	R_1 = 0.0232; wR_2 = 0.0642	R_1 = 0.0208; wR_2 = 0.0364	R_1 = 0.0197; wR_2 = 0.0372	R_1 = 0.0165; wR_2 = 0.0299	R_1 = 0.0202; wR_2 = 0.0331	R_1 = 0.0224; wR_2 = 0.0557
R indices [all data]	R_1 = 0.0282; wR_2 = 0.0667	R_1 = 0.0260; wR_2 = 0.0379	R_1 = 0.0366; wR_2 = 0.0421	R_1 = 0.0351; wR_2 = 0.0354	R_1 = 0.0264; wR_2 = 0.0343	R_1 = 0.0399; wR_2 = 0.0634
$\Delta\rho_{fin}$ (max/min), [e/Å³]	1.31/ −1.36	1.03/−1.29	1.01/−1.68	1.26/−1.05	1.09/−1.25	1.77/−1.63

All compounds are isostructural (space group: *Cmce*, № 64; Pearson's symbol-prototype: *oS*72– Ce$_2$(Ga$_{0.1}$Ge$_{0.9}$); Z = 8). From [35]

Table 4.6 Crystallographic data for Yb_2PdGe_6 single crystal taken from samples prepared by flux–synthesis [35] and of La_2CuGe_6 and La_2AgGe_6 [36] prepared by direct synthesis together with experimental details of the structural determination

Empirical formula	Yb_2PdGe_6	La_2CuGe_6	La_2AgGe_6
EDXS data	$Yb_{23.2}Pd_{10.9}Ge_{65.9}$	–	–
M_w, [g/mol]	888.02	776.90	821.23
a [Å]	8.1428(3)	8.5065(5)	8.6768(4)
b [Å]	7.9807(3)	8.1877(5)	8.3165(4)
c [Å]	21.8331(9)	21.7811(12)	21.8915(9)
V [Å^3]	1418.83(9)	1517.02(15)	1579.70(12)
Calc. density [g/cm^3]	8.314	6.803	6.906
abs coeff (μ), mm^{-1}	53.5	37.1	35.4
Unique reflections	1242	1326	1378
Reflections I > 2σ(I) R_{sigma}	1144 0.0149	840 0.0218	978 0.0120
Data/parameters	1144/50	1326/50	1378/50
GOF on F^2 (S)	1.428	1.06	1.08
R indices [I > 2σ(I)]	$R_1 = 0.0231$; $wR_2 = 0.0496$	$R_1 = 0.027$ $wR_2 = 0.039$	$R_1 = 0.015$ $wR_2 = 0.029$
R indices [all data]	$R_1 = 0.0255$; $wR_2 = 0.0506$	$R_1 = 0.039$ $wR_2 = 0.057$	$R_1 = 0.026$ $wR_2 = 0.029$
$\Delta\rho_{fin}$ (max/min), [e/Å^3]	1.92/ –1.78	1.18/–1.45	0.73/–1.12

All compounds are isostructural (space group: *Cmce*, № 64; Pearson's symbol–prototype: *oS*72–Ce$_2$(Ga$_{0.1}$Ge$_{0.9}$)$_7$; Z = 8)

the same as for the *mS*36 model. All the extra–peaks, instead, are due to the presence of the second domain, related to the first one by a 180° rotation twin law

$$: \begin{pmatrix} 1 & 0 & 0 \\ 0 & -1 & 0 \\ -\frac{1}{2} & 0 & -1 \end{pmatrix}$$

Consequently, the collected dataset was re–integrated assuming the presence of two domains and the hkl5 file for refinement was prepared by TWINABS [59]. After that, excellent residuals were obtained (see Table 4.9) for the ordered La_2PdGe_6 model with *mS*36–type structure. At the final cycles, this model was refined with anisotropic thermal displacement parameters for all atom sites. The refined fractional contribution k of the second domain is *ca.* 0.4.

The orientation of twin domains and the corresponding reciprocal plots of completely overlapped and non–overlapped *hkl* zones are shown in Fig. 4.10. Taking in mind the twin law, it becomes clear that only reflections with $h = 2n$ are affected by the twinning (*i.e.* they are completely overlapped). Considering also the presence of the only $h + k = 2n$ reflections for the C–centered lattice, it results that half of all measured intensities are affected by twinning. The intensity difference between

Table 4.7 Atomic coordinates and equivalent isotropic displacement parameters (Å²) for the studied R_2PdGe_6 single crystals with the $oS72$–$Ce_2(Ga_{0.1}Ge_{0.9})_7$ structure

Atom (site)	Atomic param.	R = Y	R = Ce	R = Pr	R = Nd	R = Er	R = Yb (f)	R = Lu
R (16g)	x/a	0.25102(5)	0.25042(2)	0.25041(2)	0.25049(2)	0.25120(2)	0.24952(3)	0.25159(3)
	y/b	0.37584(5)	0.37513(5)	0.3753(1)	0.3751(1)	0.37605(3)	0.37604(3)	0.37592(4)
	z/c	0.08125(2)	0.08270(2)	0.082540(9)	0.082326(8)	0.080953(8)	0.08257(2)	0.08075(1)
	U_{eq} (Å²)	0.0065(1)	0.004926(6)	0.00671(6)	0.00527(5)	0.005541(6)	0.00529(8)	0.0072(1)
Ge2 (8f)	x/a	0	0	0	0	0	0	0
	y/b	0.13181(8)	0.1285(1)	0.1286(2)	0.1291(3)	0.1332(1)	0.1304(1)	0.1345(1)
	z/c	0.02868(3)	0.03066(2)	0.03042(3)	0.03023(2)	0.02798(3)	0.02917(4)	0.02720(5)
	U_{eq} (Å²)	0.0073(2)	0.0063(1)	0.0084(1)	0.0065(1)	0.0063(1)	0.0062(1)	0.0086(2)
Ge3 (8f)	x/a	0	0	0	0	0	0	0
	y/b	0.11933(8)	0.1217(1)	0.1213(2)	0.1213(3)	0.1185(1)	0.1205(1)	0.1177(1)
	z/c	0.45753(3)	0.46255(2)	0.46216(3)	0.46146(2)	0.45619(3)	0.46051(4)	0.45459(5)
	U_{eq} (Å²)	0.0083(2)	0.0075(1)	0.0089(1)	0.0077(1)	0.0076(1)	0.0082(2)	0.0097(2)
Ge4 (8f)	x/a	0	0	0	0	0	0	0
	y/b	0.40519(8)	0.40516(7)	0.40475(8)	0.40492(8)	0.40524(8)	0.4047(1)	0.4052(1)
	z/c	0.19344(3)	0.19462(3)	0.19441(3)	0.19422(3)	0.19305(3)	0.19403(4)	0.19299(5)
	U_{eq} (Å²)	0.0073(2)	0.0076(1)	0.0089(2)	0.0072(2)	0.0062(1)	0.0048(1)	0.0085(2)
Ge5 (8f)	x/a	0	0	0	0	0	0	0
	y/b	0.34733(8)	0.34524(7)	0.34608(8)	0.34617(8)	0.34763(8)	0.3476(1)	0.3484(1)
	z/c	0.30752(3)	0.30553(3)	0.30572(3)	0.30599(3)	0.30810(3)	0.30685(4)	0.30862(5)
	U_{eq} (Å²)	0.0075(2)	0.0077(1)	0.0086(2)	0.0069(2)	0.0067(1)	0.0060(2)	0.0079(2)

(continued)

Table 4.7 (continued)

Atom (site)	Atomic param.	R = Y	R = Ce	R = Pr	R = Nd	R = Er	R = Yb (f)	R = Lu
Ge6 (16 g)	x/a	0.27688(6)	0.27767(5)	0.27735(4)	0.27729(4)	0.27700(5)	0.27568(7)	0.27669(8)
	y/b	0.12631(5)	0.12531(8)	0.12540(1)	0.1254(2)	0.12658(7)	0.12614(7)	0.1267(1)
	z/c	0.19308(2)	0.19461(2)	0.19442(2)	0.19416(2)	0.19259(2)	0.19379(2)	0.19237(3)
	U_{eq} (Å^2)	0.0073(1)	0.00764(8)	0.00864(9)	0.00686(8)	0.00625(9)	0.0055(1)	0.0082(2)
Pd (8f)	x/a	0	0	0	0	0	0	0
	y/b	0.12736(5)	0.12526(6)	0.12555(6)	0.12572(6)	0.12762(6)	0.12666(7)	0.12826(9)
	z/c	0.14223(2)	0.14568(2)	0.14496(2)	0.14453(2)	0.14153(2)	0.14132(2)	0.14062(3)
	U_{eq} (Å^2)	0.0060(1)	0.00664(8)	0.00765(9)	0.00585(8)	0.00504(8)	0.0039(1)	0.0068(2)

Table 4.8 Atomic coordinates and equivalent isotropic displacement parameters ($Å^2$) for La_2CuGe_6 and La_2AgGe_6 single crystals with the $oS72$–$Ce_2(Ga_{0.1}Ge_{0.9})_7$ structure [36]

La_2MGe_6 analogue	Atomic param.	La (16 g)	Ge2 (8f)	Ge3 (8f)	Ge4 (8f)	Ge5 (8f)	Ge6 (16g)	M (8f)
M = Cu	x/a	0.25068(3)	0	0	0	0	0.27872(5)	0
	y/b	0.3751(2)	0.1272(3)	0.1228(3)	0.40659(10)	0.34307(10)	0.12472(19)	0.12460(15)
	z/c	0.08372(2)	0.03256(3)	0.46280(3)	0.19421(4)	0.30552(4)	0.19460(2)	0.14804(4)
	U_{eq} ($Å^2$)	0.0050(1)	0.0065(2)	0.0074(1)	0.0084(2)	0.0081(2)	0.0085(1)	0.0100(2)
M = Ag	x/a	0.25069(2)	0	0	0	0	0.28578(3)	0
	y/b	0.37483(6)	0.12978(16)	0.12026(16)	0.41635(6)	0.33449(5)	0.12540(6)	0.12585(5)
	z/c	0.08323(2)	0.02936(2)	0.46125(2)	0.19555(2)	0.30495(2)	0.19531(2)	0.14934(2)
	U_{eq} ($Å^2$)	0.0053374(4)	0.0071(1)	0.0078(1)	0.0087(1)	0.0087(1)	0.0083(1)	0.0099(1)

Table 4.9 Crystallographic data for R_2PdGe$_6$ (R = La, Pr,) single crystals [35] taken from samples prepared by flux–synthesis and of La$_2$LiGe$_6$ [36] prepared by direct synthesis together with experimental details of the structural determination

Empirical formula	La$_2$PdGe$_6$	Pr$_2$PdGe$_6$	La$_2$LiGe$_6$
EDXS data	La$_{22.4}$Pd$_{10.5}$Ge$_{67.1}$	Pr$_{20.6}$Pd$_{11.3}$Ge$_{68.1}$	La$_{25.1}$Ge$_{74.9}$
M$_w$, [g/mol]	819.76	823.76	720.30
a [Å]	8.2163(8)	8.157(2)	8.3597(3)
b [Å]	8.4161(9)	8.328(2)	8.8595(3)
c [Å]	11.294(1)	11.195(2)	10.8539(4)
$\beta(°)$	100.477(1)	100.391(3)	100.113(1)
V [Å^3]	768.0(1)	748.1(3)	788.80(5)
Calc. density [g/cm^3]	7.090	7.314	6.07
abs coeff (μ), mm^{-1}	36.2	38.7	33.0
Twin law	[1 0 0 0 —1 0 —1/2 0 —1]		
k(BASF)	0.414(5)	0.423(5)	0.147(5)
Unique reflections	1083	1262	1403
Reflections I > 2σ(I)	1030	1062	1193
R$_{sigma}$	0.0218	0.0298	0.0171
Data/parameters	1030/50	1062/50	1403/51
GOF on F^2 (S)	1.200	1.453	1.07
R indices [I > 2σ(I)]	R$_1$ = 0.0242; wR$_2$ = 0.0773	R$_1$ = 0.0357; wR$_2$ = 0.1166	R$_1$ = 0.015; wR$_2$ = 0.033
R indices [all data]	R$_1$ = 0.0254; wR$_2$ = 0.0778	R$_1$ = 0.0501 wR$_2$ = 0.1209	R$_1$ = 0.022; wR$_2$ = 0.035
$\Delta\rho_{fin}$ (max/min), [e/Å^3]	2.33/ —2.61	4.35/ —5.29	1.85/ —0.90

All compounds are isostructural (space group: $C2/m$, № 12; Pearson's symbol–prototype: $mS36$–La$_2$AlGe$_6$; Z = 4)

overlapped/non–overlapped reflections is evident from the corresponding precession photos of $h2l$ and $h1l$ zones of reciprocal space shown in Fig. 4.10.

The crystal structure of Pr$_2$PdGe$_6$ isolated from In flux was solved in the same way as for La$_2$PdGe$_6$, just described. Its structure turned out to be monoclinic $mS36$, with the same twinning law and with similar volume ratio of the twinned domains (see Table 4.9).

Also La$_2$LiGe$_6$ turned out to be monoclinic $mS36$ with the same twinning law but a lower volume ratio between the two twinned domains (~0.15) (see Table 4.9 for more details). The lithium atoms position was easily traced as the most prominent residual peak on difference Fourier map at ~ 0.29 1/2 ~ 0.18. This position coincides with the position of Al in the structure of the La$_2$AlGe$_6$ prototype (and then with that of Pd for the two analogues synthesized here).

The atomic positions and the equivalent isotropic displacement parameters of the La$_2$PdGe$_6$, Pr$_2$PdGe$_6$ and La$_2$LiGe$_6$ monoclinic compounds are listed in Table 4.10.

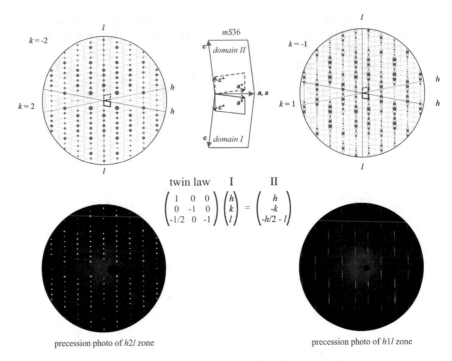

precession photo of $h2l$ zone   precession photo of $h1l$ zone

Fig. 4.10 Upper part: reciprocal orientation of twin domains together with theoretical reciprocal plots of $h2l$ (totally overlapped) and $h1l$ (non–overlapped) zones generated by XPREP [59]. Colors of domain I (red) and II (green) are the same of the corresponding hkl reflections. For clarity, relations between direct/reciprocal lattice vectors lengths are not respected. Lower part: experimental precession photos of $h2l$ and $h1l$ zones, from [35]

Positions were standardized, also in these cases, by means of the Structure Tidy program [55]. Interatomic distances are available in the Appendix (Table A4.6).

All the generated CIF files have been deposited at Fachinformationszentrum Karlsruhe, 76,344 Eggenstein–Leopoldshafen, Germany, with the following depository numbers: CSD–433026 (Y_2PdGe_6), CSD–433022 (La_2PdGe_6, flux), CSD–433081 (Ce_2PdGe_6), CSD–433025 (Pr_2PdGe_6), CSD–433205(Pr_2PdGe_6, flux), CSD–433024 (Nd_2PdGe_6), CSD–433021 (Er_2PdGe_6), CSD–433151 (Yb_2PdGe_6, flux), CSD–433023 (Lu_2PdGe_6), CSD–1871553 (La_2AgGe_6), CSD–1871552 (La_2CuGe_6) and CCSD–1871551 (La_2LiGe_6).

Experimental results reveal that the investigated 2:1:6 stoichiometry phases crystallize with the $oS72$ or the $mS36$ structures. In both cases, a complex Ge covalent framework can be highlighted (red sticks within the unit cells shown in Fig. 4.11). Analogously to what was described in paragraph 4.1, on the basis of interatomic distances analysis, it is possible to confirm the presence of ($2b$)Ge zigzag chain and ($3b$)Ge corrugated layers. Within these fragments, Ge–Ge distances range from about 2.43 to 2.60 Å, being very close to the Ge–Ge distance for αGe (2.45 Å) [1]. What is also interesting to highlight is that they are not strongly affected by the nature of

Table 4.10 Atomic coordinates and equivalent isotropic displacement parameters (Å²) for the La₂PdGe₆ and Pr₂PdGe₆ single crystals obtained by flux synthesis and for La₂LiGe₆ [36] obtained by direct synthesis

R_2MGe₆ analogue	Atomic param.	R (8j)	Ge2 (4i)	Ge3 (4i)	Ge4 (4i)	Ge5 (4i)	Ge6 (8j)	M (4i)
R = La M = Pd	x/a	0.0840(1)	0.1438(3)	0.3585(3)	0.0577(2)	0.4970(2)	0.2773(3)	0.1985(2)
	y/b	0.24980(6)	0	0	0	0	0.22206(8)	0
	z/c	0.33354(3)	0.5622(1)	0.4281(1)	0.1097(1)	0.1093(1)	0.10943(6)	0.79356(6)
	U$_{eq}$ (Å²)	0.0045(1)	0.0058(2)	0.0071(2)	0.00092(4)	0.0090(4)	0.0083(2)	0.0064(2)
R = Pr M = Pd	x/a	0.0832(3)	0.1453(6)	0.3580(6)	0.0573(3)	0.4981(3)	0.2778(4)	0.1979(2)
	y/b	0.24962(7)	0	0	0	0	0.2227(1)	0
	z/c	0.33491(4)	0.5609(2)	0.4243(1)	0.1115(2)	0.1112(2)	0.11130(9)	0.7899(1)
	U$_{eq}$ (Å²)	0.0047(2)	0.0057(3)	0.0075(3)	0.0068(5)	0.0070(5)	0.0066(2)	0.0056(2)
R = La M = Li	x/a	0.08314(2)	0.14376(6)	0.35721(6)	0.48124(6)	0.07539(6)	0.27800(5)	0.2048(11)
	y/b	0.24935(2)	0	0	0	0	0.21357(4)	0
	z/c	0.33174(2)	0.56790(4)	0.42301(4)	0.11284(5)	0.11281(5)	0.11223(3)	0.81877(8)
	U$_{eq}$ (Å²)	0.0059(1)	0.0076(1)	0.0079(1)	0.011(1)	0.011(1)	0.011(1)	0.015(2)

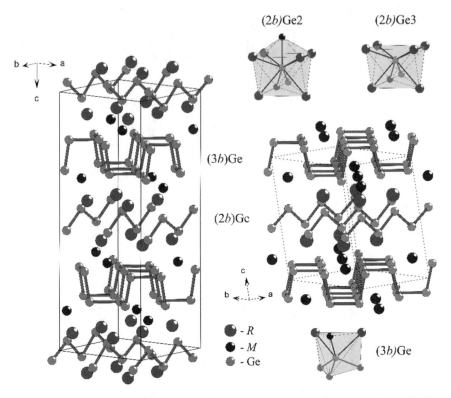

Fig. 4.11 The two structural modifications of the R_2MGe_6 compounds ($R = $ rare earth metal; $M = $ other metal): the $oS72$–$Ce_2(Ga_{0.1}Ge_{0.9})_7$ (left) and the $mS36$–La_2AlGe_6 (right). Ge covalent fragment, deduced on the basis of interatomic distances, are represented together with Ge coordination polyhedra

the lanthanide elements. There are no reference compounds (like αGe for Ge–Ge distances) that allow to describe chemical bonding on the basis of Ge–M, Ge–R and R–M distances. Within the structures, the $(2b)$Ge were labeled in Tables 4.7, 4.8 and 4.10 as Ge2 and Ge3 and show two different coordination polyhedra: a capped trigonal prism for $(2b)$Ge2 and a trigonal prism for $(2b)$Ge3 (Fig. 4.11). The additional capping position is due to a $(2b)$Ge2–M contact. On the other hand, all the $(3b)$Ge, corresponding to Ge4, Ge5 and Ge6, have the same strongly distorted trigonal prismatic coordination (Fig. 4.11).

Focusing on the R_2PdGe_6 series, the unit cell volumes were plotted as a function of the trivalent rare earth metal radii [60] (Fig. 4.12). The monoclinic cell volume was doubled in order to compare the different related structures [14]. Only lattice parameters obtained from single crystal X–ray diffraction were considered.

In agreement with the lanthanide contraction, a linear decreasing trend is observed, suggesting a similar chemical bonding scenario for all the studied germanides. Literature data on Dy_2PdGe_6 [15] fit quite well in the general trend. The most significant deviation is observed for Yb_2PdGe_6; this result is in good agreement with a

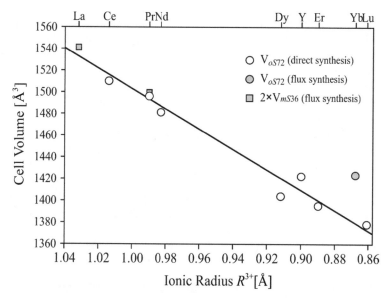

Fig. 4.12 Normalized cell volumes (from single crystal data) of R_2PdGe$_6$ compounds as a function of the R^{3+} ionic radius, from [35]. The cell volume of Dy$_2$PdGe$_6$ was taken after [15]

recent magnetic investigation [8] revealing for Yb a behavior typical of dynamic intermediate valence systems.

The crystal structure of the R_2PdGe$_6$ (R = Sm, Gd, Tb, Dy, Ho, Tm) compounds was not studied by single crystal X–ray diffraction. Powder diffraction patterns, recorded on the corresponding samples, can be satisfactorily indexed with both $oS72$ and $mS36$ structures. However, considering results obtained on the directly synthesized samples with early and late rare earth metals, it is reasonable to suggest that in the same conditions even the abovementioned R_2PdGe$_6$ phases are $oS72$–Ce$_2$(Ga$_{0.1}$Ge$_{0.9}$)$_7$ type.

A similar rationalization could not be done for the crystal structure of the 2:1:6 germanides synthesized by flux method since for Pr$_2$PdGe$_6$ a different structure was stabilized, instead for Yb$_2$PdGe$_6$ the same orthorhombic modification forms.

4.5 From R_2PdGe$_6$ to R_2MGe$_6$ Compounds: Crystallochemical and DFT Analyses Targeting the Correct Structural Model[5]

From results described above, it is clear that the $oS18$–Ce$_2$CuGe$_6$ structure never realizes in the R_2PdGe$_6$ series. Within this family, direct synthesis always produces compounds belonging to the $oS72$–Ce$_2$(Ga$_{0.1}$Ge$_{0.9}$)$_7$ structure type, except

---

[5]Part of this section was originally published in [35].

La_2PdGe_6, whose structure in this conditions was not definitely confirmed. The In–flux synthetic route applied in this study gave origin to $mS36$–La_2AlGe_6 twinned crystals of La_2PdGe_6 and Pr_2PdGe_6 whereas Yb_2PdGe_6 crystals grown from the same metal solvent are orthorhombic $oS72$.

These results are coherent with recent structural studies on 2:1:6 germanides assigning to them the $oS72$ [6, 8, 15, 16] or the $mS36$ model [18] instead of the previously proposed $oS18$.

In view of the following discussion it is convenient to distribute all the known 2:1:6 models within a concise scheme (Bärnighausen tree) starting from the $oS20$–$SmNiGe_3$ aristotype. When considering the vacancy ordering phenomenon [17, 61, 62] causing the symmetry reduction, each structural model finds its location on a separate branch. To capture the changes taking place, it is sufficient to follow the M–Ge sub–lattice distortions as emphasised in Fig. 4.13 (R atom positions are negligibly affected by the symmetry reduction).

From the crystal chemical point of view, as anticipated in paragraph 4.1, the $oS18$ model is somewhat suspicious:

– it contains many independent crystallographic Ge sites, not in agreement with the symmetry principle [63];
– the inhomogeneous vacancies distribution implies that corrugated Ge layers are linked to the zig–zag germanium chains through bridging M atoms only on one side (see Fig. 4.13).

To corroborate this idea arising from symmetry considerations, the DFT total energy values were calculated for La_2PdGe_6 and Y_2PdGe_6 with the $oS18$, $oS72$ and $mS36$ structures. Structural data available in the literature or obtained during this work were used as starting point for geometric relaxation. In the case of Y_2PdGe_6, for which the $mS36$ model was never reported, cell dimensions and atomic positions of the La–analogue were chosen. The choice of these rare earths permits to avoid the presence of highly correlated electrons lying in partially filled f states. The relaxed structures, obtained at the end of the variable–cell calculations, showed lattice constants bigger of about 2% with respect to literature data and results presented in this work, as expected when using PBE functional (see Table 4.11).

On the basis of obtained results the $oS18$ structural model is the worst: its total energy is higher by 0.063 eV/atom for La_2PdGe_6 and by 0.066 eV/atom in the case of Y_2PdGe_6 with respect to both $oS72$ and $mS36$ models.

Calculations for the $oS18$ and $oS72$ models were performed on other La_2MGe_6 analogues (M = Pt, Cu, Ag, Au). As can be seen from the plot shown in Fig. 4.14, the highest energy value corresponds always to the $oS18$ model ($\Delta E(eV/at)$ = $E_{oS72} - E_{oS18}$, is always negative).

The same was also done for La_2LiGe_6 where the energy difference between the hypothetical $oS72$ and $oS18$ (obviously with Pr_2LiGe_6 structure type and not Ce_2CuGe_6) turned out to be –0.076 eV/at.

The aforementioned experiments on Li, Cu and Ag analogues were performed after these results were obtained in order to check the validity of the complete structure revision here proposed on the basis of calculated total energies. It is important

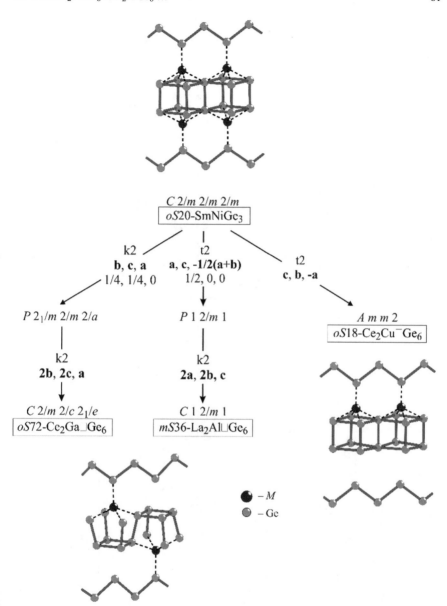

Fig. 4.13 Bärnighausen tree relating the SmNiGe$_3$ aristotype and its orthorhombic and monoclinic vacancy variants. The type and indexes of the symmetry reductions are indicated. For clarity, only M–Ge framework are shown for each structural model; Ge–Ge contacts are shown in red, M–Ge interactions by a dotted black line. From [35]

Table 4.11 Experimental and calculated parameters for R_2PdGe_6 (R = La, Y) in the $oS18$, $oS72$ and $mS36$ modifications

Compound		Experimental			Calculated		
		$oS18$	$oS72$	$mS36$	$oS18$	$oS72$	$mS36$
Y_2PdGe_6	a (Å)	4.0790(4)	8.1703(5)		4.1502	8.3117	8.2181
	b (Å)	4.0168(5)	8.0451(5)		4.0802	8.2239	8.3140
	c (Å)	21.525(2)	21.558(2)		22.212	21.857	11.123
	β(°)						100.5
	V (Å³)	352.7(1)	1417.1(2)		376.14	1494.0	747.16
	Energy (eV/at)				− 691.647	− 691.713	− 691.713
	Ref	2					
La_2PdGe_6	a (Å)	4.2117(3)	8.430(1)	8.2163(8)	4.2914	8.5552	8.3910
	b (Å)	4.1100(3)	8.2180(7)	8.4161(9)	4.1893	8.3902	8.5554
	c (Å)	22.265(5)	22.192(3)	11.294(1)	22.913	22.642	11.515
	β(°)			100.5(1)			100.5
	V (Å³)	385.4(1)	1537.4(4)	768.0(1)	411.92	1625.2	812.81
	Energy (eV/at)				− 721.181	− 721.244	− 721.244
	Ref	2	3				

If not specified, data were obtained in this work. From [35]

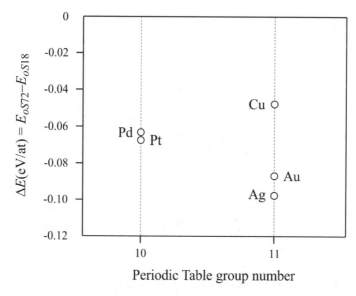

Fig. 4.14 ΔE versus nature of the late transition element M for La_2MGe_6. From [35]

to underline that the $mS36$ model was not considered for energy calculations of La_2MGe_6 because results on La_2PdGe_6 and Y_2PdGe_6 (Table 4.11) do not show any energy difference between $oS72$ and $mS36$. Anyway, after the crystal structure solution of the La_2LiGe_6 phase which turned out to be monoclinic, one more calculation was performed according to this structure. Again, no differences were obtained between the $oS72$ and $mS36$ modifications. In fact, they can be considered as two different polytypes of 2:1:6 composition. They both consist of geometrically equivalent layers of defective $BaAl_4$ (of $R\square MGe_2$ composition), AlB_2 and α–Po type slabs stacked linearly along the c direction [14, 19]. These layers are also energetically equivalent, and consequently no strong preference for one of the polytypes can be envisaged. Frequently, the structure of such compounds is sensitive to the crystallization conditions and small fluctuations of these may reverse the energetic preferences and at the same time give origin to stacking faults, twinnings or non–periodic structures; as an example the families of SiC, ZnS, CdI_2, micas, etc. can be cited [64].

These findings do not preclude the possibility that new polytypes exist for this numerous group of compounds. One can suppose that modulated structures may form within this family. The non–periodicity could arise from a vacancy ordering or Ge covalent fragments distortions with a periodicity different from that of lattice.

Some more aspects associated with the symmetry reduction scheme can be highlighted. The symmetry reduction path conducting to the $mS36$ model contains a *translationengleiche* transformation of index 2 that matches perfectly the fact that non–merohedral two domain twins form. The presence of *klassengleiche* relations, instead, should be at the origin of antiphase domains in both $oS72$ and $mS36$ polytypes.

These domains are not detectable by conventional X–ray diffraction techniques; however, the quality and dimensions of single crystals obtained by flux method are well promising for further transmission electron microscopic investigations targeting this goal.

It remains, however, unclear if the $oS20$ aristotype has any physical meaning. Further investigations should be performed aiming to clarify if elevated temperature/pressure conditions may stabilize the hypothetical $oS20$–RPdGe$_3$. If yes, the found non–merohedral twins have been developed by a phase transition in the solid state. Otherwise, their formation took place during the growth of the crystal.

4.6   Electronic Structure and Chemical Bonding Analysis for La$_2M$Ge$_6$ Phases ($M =$ Li, Mg, Al, Zn, Cu, Ag, Pd) and Y$_2$PdGe$_6$[6]

Calculated DOS curves (Fig. 4.15) for the title compounds show metal–like behavior, as expected from considerations reported in the first paragraph, with pseudo–gap at the Fermi level E_F. The main contributions are due to Ge derived states: the $4s$ ones essentially dominate the region below -5 eV and the $4p$ ones mainly contribute in the range from about -5 to 0 eV. The M states are dispersed over a wide energy range whereas La $5d$ (Y $4d$) contribute essentially in the vicinity of E_F yielding to an energetic overlap with Ge p states, supporting the idea of polar heteronuclear La–Ge bonds.

For all the compounds, the locations of the local maxima (attractors) of the ELI–D spatial distribution at about the midpoint of the nearest neighbor Ge–Ge contacts (Fig. 4.17) were consistent with the picture of covalently bonded anionic zigzag chains of $(2b)$Ge and the corrugated layers of $(3b)$Ge anions derived from interatomic distances and the 8–N rule (Zintl–Klemm approach) (Fig. 4.16a).

In the Lewis picture, the number of electron pairs around each Ge species should be four, hence $(2b)$Ge and $(3b)$Ge are expected to display 2 and 1 lone pair type ELI–D attractors, respectively, which is not strictly fulfilled. In detail, for $M =$ Li, Mg, Al one additional ELI–D attractor beyond the two expected ones was observed for species $(2b)$Ge2 (Fig. 4.17). This already marks a certain deviation to be discussed below. As a further common feature for all compounds, La atoms display a significant structuring of the penultimate shell (structuring index $\varepsilon = 0.06$ [65], which signals its noticeable participation in covalent bonding (Fig. 4.17).

From the QTAIM analysis, atomic regions were obtained, and their electronic populations yield the corresponding effective atomic charges $Q^{eff}(X)$ (Tables 4.12, 4.13 and 4.14).

The charge trend between the formal $(2b)$Ge^{2-} and $(3b)$Ge$^-$ species, and the Li$^+$ and Mg^{2+} species is correctly represented by their effective charges. One may note from Fig. 4.16d, that the magnitudes of the effective charges are smaller than the

---

[6]Part of this section was originally published in [36].

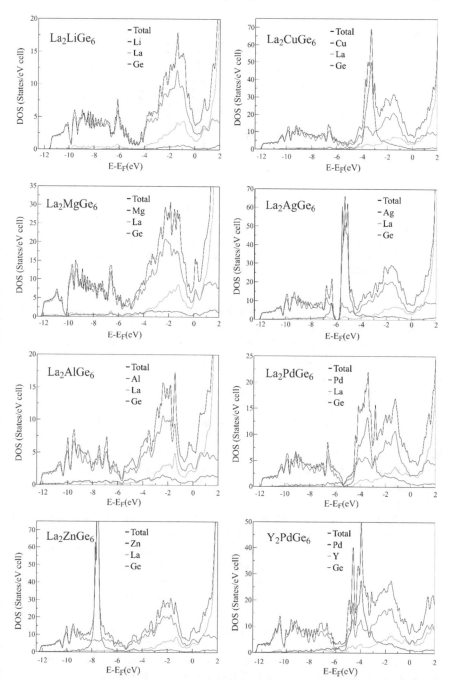

Fig. 4.15 Total and projected density of states (DOS/pDOS) for the title compounds. Adapted with permission from [36]

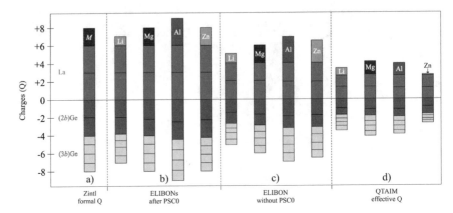

Fig. 4.16 a Formal charges Q deduced applying Zintl rule to a generic $La^{III}{}_2M^{II}Ge_6$ compound (M^{II} = divalent metal), **d** QTAIM effective charges and ELI Based Oxidation Numbers (ELIBONs) obtained (c) before and (b) after the application of the new PSC0 correction for La_2MGe_6, M = Li, Mg, Al, Zn. Adapted with permission from [36]

formal ones (Fig. 4.16a) in all cases, which is typically observed, *e.g.* [66]. Moreover, La and Al atoms display strikingly low effective charges with respect to their formal values 3+ , even lower than the Mg one. This issue was investigated by means of the QTAIM/ELI–D basin intersection technique including penultimate shell corrections (PSC0) of the basin populations.

Integration of the electron density inside the ELI–D core regions for all atomic species yields certain deviations from the ideal core electron count given by imposing the chemically active number of valence electrons (Eq. 3.38). For Ge atoms always an under population of the core region of about 0.22 e is found, which corresponds to a –0.22 e per Ge atom overpopulation of the valence region (Eq. 3.39). In contrast, for the metal atoms Li, Mg, Al, La a core overpopulation and valence region under population is observed, while Zn behaves qualitatively like Ge. The corresponding values are given in Table 4.15

To a large extent, these effects already occur for the free atoms [67, 68] and are affected by the bonding situation only by a virtually negligible degree. With La displaying the largest deviation of about 1 electron valence under population, the correction of all these deviations from chemical valence using the PSC0 approximation turns out to be decisive for the classification of the polarity of Ge–La bonds in the studied compounds. It is important to note, that the QTAIM effective charges are not dependent on ELI–D shell structure, and the applied PSC0 correction therefore does not change them. The correction mechanism works on the basis of the ELI–D/QTAIM basin intersection, and the population correction of an ELI–D valence basin occurs in that part of the intersected basin, which belongs to the QTAIM atom whose ELI–D valence population is to be corrected (Eqs. 3.39, 3.40 and Fig. 3.5). Since in the ELIBON formalism all valence basins of the access set s_{Ge} of each Ge atom are specifically attributed to this Ge atom, after PSC0 correction the ELIBON of the metal atoms display the value consistent with the chemical valence, while before

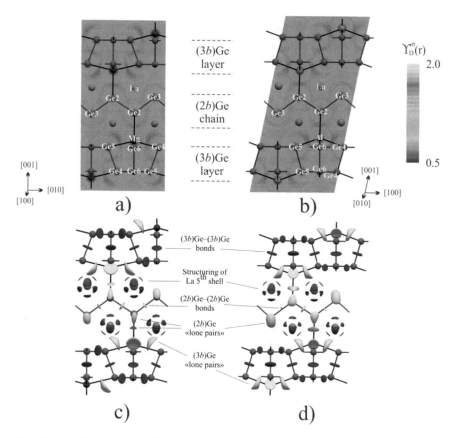

Fig. 4.17 ELI–D distribution within the La$_2$MgGe$_6$ (**a** and **c**) and the La$_2$AlGe$_6$ (**b** and **d**) crystal structures, selected as a representative for the *oS*72 and *mS*36 phases, respectively. The same isosruface colours (**c** and **d**) were chosen to highlight topologically similar attractors: (2*b*)Ge–(2*b*)Ge bonds white, (3*b*)Ge–(3*b*)Ge bonds blue, Ge "lone–pairs" yellow and La penultimate shell red

this was clearly not the case (Fig. 4.16b, c). It is instructive to see, how close the corrected ELIBON of the different Ge species for La$_2$MGe$_6$ (M = Mg, Zn) comes to the formal values for the ideal 8–*N* electron count (Fig. 4.16a, b). The deviation of the ELIBON picture from the effective charges (Fig. 4.16d) is caused by polar interactions of the Ge polyanionic partial structure with the metal atoms, which is not included in the classical 8–*N* framework.

As already noted, the Ge–Ge' interactions of the classical model were all found to be signified by a corresponding ELI–D attractor and its associated basin. The participation of the metal atoms in these bond basins as measured by the corresponding bond fraction (Eq. 3.22) $p(\text{bond}^M)$ of metal *M* (see Tables A4.7–4.10) was always well below 0.1. This is markedly smaller than the effects exhibited by the lone pair type regions on which the focus will be set in the following. As a measure of these

Table 4.12 Position–space bonding analysis for La$_2$LiGe$_6$, La$_2$MgGe$_6$ and La$_2$AlGe$_6$

Without PSC0

Atoms (Ω)	Q^{form}	$Q^{eff \cdot}(\Omega)$	ELIBON	N_{val}^{ELI} (Ge)	N_{acc}^{ELI} (C^{Ge})	N_{cb}(Ge)	N_{lp}(Ge)
La	+ 3	+ 1.3183	+ 2.04740	–	–	–	–
(2b)Ge2	–	−0.9684	−1.39450	5.22450	7.05340	1.78420	3.44030
(2b)Ge3	–	−0.7870	−1.25295	5.04870	6.75380	1.60160	3.44710
(3b)Ge4	–	−0.4027	−0.53790	4.65750	7.58430	2.92650	1.73100
(3b)Ge5	–	−0.4142	−0.55105	4.66930	7.57620	2.90570	1.76360
(3b)Ge6	–	−0.4422	−0.65705	4.70260	7.71120	2.88770	1.81490
Li	+ 1	+ 0.8199	+ 0.95720	–	–	–	–
La	+ 3	+ 1.3272	+ 2.0395	–	–	–	–
(2b)Ge2	−2	−1.1102	−1.59430	5.36190	7.09750	1.70320	3.65870
(2b)Ge3	−2	−0.7814	−1.25400	5.03840	6.59970	1.45710	3.58130
(3b)Ge4	−1	−0.5536	−0.76015	4.81280	7.85500	3.03250	1.78030
(3b)Ge5	−1	−0.5554	−0.75965	4.81300	7.85300	3.03730	1.77570
(3b)Ge6	−1	−0.5630	−0.81420	4.82330	7.93380	3.00890	1.81440
Mg	+ 2	+ 1.4715	+ 1.9263	–	–	–	–
La	+ 3	+ 1.2884	+ 2.0075	–	–	–	–
(2b)Ge2	–	−1.1641	−1.87090	5.42480	7.32620	1.86590	3.55890
(2b)Ge3	–	−0.7723	−1.29410	5.02790	6.64540	1.47150	3.55640
(3b)Ge4	–	−0.5290	−0.95250	4.78330	8.17220	3.38880	1.39450
(3b)Ge5	–	−0.5282	−0.96910	4.77880	8.18300	3.40420	1.37460
(3b)Ge6	–	−0.4750	−0.92865	4.72870	8.19180	3.29780	1.43090
Al	+ 3	+ 1.3668	+ 2.9254	–	–	–	–

After PSC0

La	+ 3	+ 1.3183	+ 3.00000	–	–	–	–
(2b)Ge2	–	−0.9684	−1.98728	4.95740	7.35240	2.26699	2.69041
(2b)Ge3	–	−0.7870	−1.82017	4.78261	7.02259	2.06243	2.72018
(3b)Ge4	–	−0.4027	−0.72731	4.39750	7.37102	2.94555	1.45195
(3b)Ge5	–	−0.4142	−0.74813	4.40850	7.38529	2.92305	1.48545
(3b)Ge6	–	−0.4422	−0.85792	4.43659	7.50888	2.89251	1.54408
Li	+ 1	+ 0.8199	+ 1.00000	–	–	–	–
La	+ 3	+ 1.3272	+ 3.00000	–	–	–	–
(2b)Ge2	−2	−1.1102	−2.20571	5.09730	7.41421	2.23774	2.85956
(2b)Ge3	−2	−0.7814	−1.81685	4.77570	6.86708	1.93220	2.84350
(3b)Ge4	−1	−0.5536	−0.96134	4.54841	7.65600	3.03800	1.51041
(3b)Ge5	−1	−0.5554	−0.96319	4.54960	7.65832	3.04796	1.50164
(3b)Ge6	−1	−0.5630	−1.02209	4.55601	7.73628	3.01397	1.54204
Mg	+ 2	+ 1.4715	+ 2.00000	–	–	–	–
La	+ 3	+ 1.2884	+ 3.00000	–	–	–	–
(2b)Ge2	–	−1.1641	−2.51375	5.15841	7.67475	2.42527	2.73314
(2b)Ge3	–	−0.7723	−1.90107	4.76271	6.95666	1.98836	2.77435
(3b)Ge4	–	−0.5290	−1.15010	4.52495	7.97858	3.39853	1.12642
(3b)Ge5	–	−0.5282	−1.17138	4.52180	8.01158	3.39013	1.13167
(3b)Ge6	–	−0.4750	−1.13347	4.46949	8.00938	3.28260	1.18689
Al	+ 3	+ 1.3668	+ 3.00000	–	–	–	–

Values before and after PSC0 correction, applied choosing atomic core charges corresponding to Ge^{4+}, Li^{+}, Mg^{2+}, Al^{3+} and La^{3+}, are given. Quantities referred to each species are reported. The values after PSC0 correction are reproduced with permission from [36]

Table 4.13 Position–space bonding analysis for La_2ZnGe_6 and La_2CuGe_6

Without PSC0

Atoms (Ω)	Q^{form}	$Q^{eff.}$ (Ω)	ELIBON	N_{val}^{ELI} (Ge)	N_{acc}^{ELI} (C^{Ge})	N_{cb}(Ge)	N_{lp}(Ge)
La	+3	+1.2970	+2.0129	–	–	–	–
(2b)Ge2	−2	−0.8556	−1.78610	5.1188	7.4349	2.2750	2.8438
(2b)Ge3	−2	−0.7691	−1.29415	5.0298	6.7722	1.5805	3.4493
(3b)Ge4	−1	−0.2528	−0.87980	4.5047	8.1066	3.5635	0.9412
(3b)Ge5	−1	−0.2528	−0.88235	4.5039	8.1113	3.5783	0.9256
(3b)Ge6	−1	−0.2803	−0.83805	4.5341	8.0869	3.4529	1.0812
Zn	+2	+0.0946	+2.4949	–	–	–	–
La	+3	+1.2859	+2.00350	–	–	–	–
(2b)Ge2	−2	−0.7818	−1.68780	5.0423	7.4777	2.3871	2.6552
(2b)Ge3	−2	−0.7482	−1.27595	5.0561	6.9658	1 0.7184	3.3377
(3b)Ge4	−1	−0.2047	−0.82730	4.4473	8.0462	3.5107	0.9366
(3b)Ge5	−1	−0.1620	−0.78945	4.4482	8.0494	3.5209	0.9273
(3b)Ge6	−1	−0.2025	−0.74485	4.4763	8.0055	3.4693	1.0070
Cu	+2	−0.2059	+2.13810	–	–	–	–

After PSC0

La	+3	+1.2970	+3.0000	–	–	–	–
(2b)Ge2	−2	−0.8556	−2.29637	4.85259	7.65199	2.69760	2.15499
(2b)Ge3	−2	−0.7691	−1.89305	4.76320	7.07260	2.07622	2.68698
(3b)Ge4	−1	−0.2528	−0.96001	4.24910	7.80469	3.42197	0.82713
(3b)Ge5	−1	−0.2528	−0.96328	4.24770	7.80990	3.43584	0.81186
(3b)Ge6	−1	−0.2803	−0.94257	4.27611	7.80526	3.33742	0.93869
Zn	+2	+0.0946	+2.0000	–	–	–	–
La	+3	+1.2859	+3.00000	–	–	–	–
(2b)Ge2	–	−0.7818	−2.05029	4.78000	7.55664	2.65714	2.12286
(2b)Ge3	–	−0.7482	−1.90164	4.73500	7.23472	2.20003	2.53497
(3b)Ge4	–	−0.2047	−0.76903	4.20190	7.61697	3.23506	0.96684
(3b)Ge5	–	−0.1620	−0.74658	4.15881	7.58289	3.22061	0.93820
(3b)Ge6	–	−0.2025	−0.72888	4.19949	7.57962	3.21050	0.98899
Cu	+1	−0.2059	+1.00000	–	–	–	–
La	+3	+1.2859	+3.00000	–	–	–	–
(2b)Ge2	−2	−0.7818	−2.26562	4.78000	7.77197	2.87247	1.90753
(2b)Ge3	−2	−0.7482	−1.90164	4.73500	7.23472	2.20003	2.53497
(3b)Ge4	−1	−0.2047	−0.97757	4.20190	7.82551	3.44360	0.75830
(3b)Ge5	−1	−0.1620	−0.96016	4.15881	7.79647	3.43419	0.72462
(3b)Ge6	−1	−0.2025	−0.91014	4.19949	7.76089	3.39177	0.80772
Cu	+2	−0.2059	+2.00000	–	–	–	–

Values before and after PSC0 correction, applied choosing atomic core charges corresponding to Ge^{4+}, La^{3+}, Zn^{2+}, Cu^{+} and Cu^{2+} are given. Quantities referred to each species are reported. The values after PSC0 correction are reproduced with permission from [36]

Table 4.14 Position–space chemical bonding analysis for La$_2$AgGe$_6$, La$_2$PdGe$_6$ and Y$_2$PdGe$_6$

Atoms (Ω)	Q^{form}	$Q^{eff.}(\Omega)$	ELIBON	N_{val}^{ELI} (Ge)	N_{acc}^{ELI} (C^{Ge})	N_{cb}(Ge)	N_{lp}(Ge)
Without PSC0							
La	+3	+1.3295	+2.0281	–	–	–	–
(2*b*)Ge2	–	−0.7581	−1.52930	5.0192	7.2632	2.1977	2.8215
(2*b*)Ge3	–	−0.7914	−1.31740	5.0557	6.8613	1.6364	3.4193
(3*b*)Ge4	–	−0.1602	−0.67120	4.4108	7.9444	3.4900	0.9208
(3*b*)Ge5	–	−0.1719	−0.68465	4.4208	7.9518	3.4887	0.9321
(3*b*)Ge6	–	−0.2079	−0.68075	4.4601	7.9532	3.4058	1.0543
Ag	+1	−0.3631	+1.5063	–	–	–	–
La	+3	+1.3381	+2.0362	–	–	–	–
(2*b*)Ge2	−2	−0.6749	−1.45405	4.9349	7.2911	2.3019	2.6330
(2*b*)Ge3	−2	−0.8202	−1.37910	5.0930	7.0880	1.7626	3.3304
(3*b*)Ge4	−1	−0.1140	−0.72520	4.3574	8.0253	3.5779	0.7795
(3*b*)Ge5	−1	−0.1099	−0.72970	4.3535	8.0296	3.5797	0.7738
(3*b*)Ge6	−1	−0.1344	−0.67740	4.3795	7.9817	3.5500	0.8295
Pd	+2	−0.6889	+1.5733	–	–	–	–
Y	+3	+1.5384	+2.2394	–	–	–	–
(2*b*)Ge2	−2	−0.7766	−1.7293	5.0270	7.4777	2.3966	2.6304
(2*b*)Ge3	−2	−0.9341	−1.4825	5.1912	6.7602	1.4104	3.7808
(3*b*)Ge4	−1	−0.1196	−0.7518	4.3624	8.0971	3.6320	0.7304
(3*b*)Ge5	−1	−0.1332	−0.7754	4.3754	8.1261	3.6745	0.7009
(3*b*)Ge6	−1	−0.1393	−0.73150	4.3811	8.0875	3.6268	0.7543
Pd	+2	−0.8356	+1.7228	–	–	–	–
After PSC0							
Y	+3	+1.5384	+3.00000	–	–	–	–
(2*b*)Ge2	−2	−0.7766	−2.19074	4.76969	7.64745	2.76165	2.00804
(2*b*)Ge3	−2	−0.9341	−1.96420	4.92409	6.93444	1.81725	3.10684
(3*b*)Ge4	−1	−0.1196	−0.87845	4.11630	7.85327	3.53809	0.57821
(3*b*)Ge5	−1	−0.1332	−0.91227	4.12990	7.89392	3.58878	0.54112
(3*b*)Ge6	−1	−0.1393	−0.85732	4.13440	7.84190	3.53027	0.60413
Pd	+2	−0.8356	+1.66008	–	–	–	–

Values after PSC0 correction, applied choosing atomic core charges corresponding to Ge^{4+}, Y^{3+}, Pd^{2+}, are given only for Y$_2$PdGe$_6$. Quantities referred to each species are reported

polar interactions the bond fractions $p(\text{lp}^{Ge})$ of the Ge lone pair type ELI–D basins are employed, which are equal to 1 in the idealized framework meaning 100% lone pair character (lpc(lp) = 1, Eq. 3.23). As can be seen in Table 4.16, the deviations from the ideal case are significant for all lone pair type basins already for the hard cations $M = $ Li$^+$, Mg^{2+}. For example, assuming a Zintl scenario, the bonding pattern for Ge3 species should be (2b, 2lp); with Ge–MLa$_n$ interactions being not purely ionic but polar (p), the pattern is (2b, 2p). Since La atoms are present in the access set, *i.e.* in the coordination sphere, of all Ge lone pair type basins, Ge–La polar–covalent bonding dominates the lone pair characters in all these cases. The large degree of Ge–La covalency becomes evident for Ge2 and Ge3 lone pair type regions after the PSC0 correction (see Tables A4.6–A4.10). It is interesting to note that for Ge3 each

Table 4.15 ELI–D core basins electronic population for each species within the unit cell of La_2MGe_6 M = Li, Mg, Al, Zn and the valence shell defects (Eq. 3.39) are reported

Compound	Species X	N_{core}	$\sum_{i=1}^{n-1} N_i^{ELI}(X)$	Lit [67, 68]	$N_{vdef}^{ELI}(X)$
La_2LiGe_6	La	54	54.9526	54.726	− 0.9526
	(2b)Ge2	28	27.7329	27.735	+ 0.2671
	(2b)Ge3		27.7339		+ 0.2661
	(3b)Ge4		27.7400		+ 0.2600
	(3b)Ge5		27.7392		+ 0.2608
	(3b)Ge6		27.7340		+ 0.2660
	Li	2	2.0428	2.0	− 0.0428
La_2MgGe_6	La	54	54.9605	54.726	− 0.9605
	(2b)Ge2	28	27.7354	27.735	+ 0.2646
	(2b)Ge3		27.7372		+ 0.2628
	(3b)Ge4		27.7356		+ 0.2644
	(3b)Ge5		27.7366		+ 0.2634
	(3b)Ge6		27.7327		+ 0.2673
	Mg	10	10.0737	10.1	− 0.0737
La_2AlGe_6	La	54	54.9925	54.726	− 0.9925
	(2b)Ge2	28	27.7336	27.735	+ 0.2664
	(2b)Ge3		27.7348		+ 0.2652
	(3b)Ge4		27.7416		+ 0.2584
	(3b)Ge5		27.7430		+ 0.2570
	(3b)Ge6		27.7408		+ 0.2592
	Al	10	10.0746	10.1	− 0.0746
La_2ZnGe_6	La	54	54.9871	54.726	− 0.9871
	(2b)Ge2	28	27.7338	27.735	+ 0.2662
	(2b)Ge3		27.7334		+ 0.2666
	(3b)Ge4		27.7444		+ 0.2556
	(3b)Ge5		27.7438		+ 0.2562
	(3b)Ge6		27.7420		+ 0.2580
	Zn	28	27.5051	27.426	+ 0.4949

Adapted with permission from [36]

lone pair type basin is only surrounded (access set) by 3 La atoms, so that its covalent character is caused solely by La coordination and does not virtually change with the type of M atom (Table 4.16).

For Ge2 species there also exist two lone pair type regions, which do not have M atoms in their access set (M = Li, Mg, Al), and consequently display very similar lone pair character for all these compounds (Tables A4.7–A4.9).

Table 4.16 Bonding characteristics for La_2LiGe_6, La_2MgGe_6, La_2AlGe_6, and La_2ZnGe_6 after PSC0 correction

	Li	Mg	Al	Zn
	Ge2	Ge2	Ge2	Ge2
2b	(0.67b)	(0.61b)	(0.61b)	(0.69b),
	(0.63b)	(0.55b)	(0.48b', 0.03lp')	(0.60b', 0.02lp')
2p	(0.36b', 0.41lp')	(0.39b', 0.44lp')	(0.38b', 0.44lp')	(0.70b', 0.53lp')
	(0.36b', 0.41lp')	(0.39b', 0.44lp')	(0.38b', 0.44lp')	(0.70b', 0.53lp')
1p	(0.24b', 0.53lp')	(0.29b', 0.54lp')	(0.57b', 0.46lp')	–
	Ge3	Ge3	Ge3	Ge3
2b	(0.53b), (0.59b)	(0.47b), (0.50b)	(0.52b), (0.43b)	(0.53b), (0.53b)
2p	(0.47b', 0.68lp')	(0.48b', 0.71lp')	(0.52b', 0.69lp')	(0.51b', 0.67lp')
	(0.47b', 0.68lp')	(0.48b', 0.71lp')	(0.52b', 0.69lp')	(0.51b', 0.67lp')
	Ge4	Ge4	Ge4	Ge4
3b	2(0.85b)	2(0.91b)	2(0.94b)	2(0.90b)
	(0.92b)	(0.85b)	(0.91b)	(0.98b)
1p	(0.32b', 0.73lp')	(0.38b', 0.76lp')	(0.60b', 0.56lp')	(0.64b', 0.41lp')
	Ge5	Ge5	Ge5	Ge5
3b	2(0.85b)	2(0.91b)	3(0.93b)	2(0.90b)
	(0.92b)	(0.85b)		(0.98b)
1p	(0.31b', 0.74lp')	(0.38b', 0.75lp')	(0.60b', 0.57lp')	(0.65b', 0.40lp')
	Ge6	Ge6	Ge6	Ge6
3b	2(0.81b)	2(0.88b)	2(0.89b)	2(0.89b)
	(0.93b)	(0.87b)	(0.96b)	(0.99b)
1p	(0.33b', 0.77lp')	(0.38b', 0.77lp')	(0.54b', 0.59lp')	(0.57b', 0.47lp')

All classically expected Ge lone pairs form polar covalent Ge–MLa$_n$ polyatomic bonds p = x b' + y lp', which are decomposed into covalent b' and polar lp' (hidden lone pairs) contributions, respectively. Symbols b, b', and lp' correspond to two–electron bond and lone pair type contributions; For M = Al, Zn the covalent bond Ge2–Ge3 was found to display a tiny polar contribution as well. Reproduced with permission from [36]

As a special ELI–D feature, (2*b*)Ge2 species displays five ELI–D attractors for M = Li, Mg, Al, instead of the four expected ones. The unexpected basin is easily identified being the one not consistent with a pseudo tetrahedral arrangement of Ge–Ge bonds and lone pairs. It is the one with an ELI–D attractor along the line Ge–M. Instead of the attractor a (3, −1) saddle point of the ELI–D is expected at this position being virtually the interconnection point between the two classical lone pair regions. Thus, the topology at this point is characterized by a change of sign of the ELI–D curvature along the M–Ge line from positive (expected) to negative. The additional lone pair type region has evolved at the cost of the two classically expected ones, as can be also seen comparing the sum of the three lone pair type basin occupations for Ge2 species with the two ones for Ge3 species being 4.3 e rather similar in both cases. The strongly polarizing cations of Li, Mg, and Al at close distance seem to be the primary cause for this extra splitting, but, as can be seen in Table 4.16 and A4.7–A4.9, both, the two classical lone pair regions and the additional one display a rather high polar–covalent character due to the participating three and four La atoms, respectively. As a result of this "softening" of the lone pair character towards a multiatomic bonding feature, the "resistance" of initially two classical lone pair type regions against splitting into three ones might have become decreased. It is important to distinguish between the obvious influence of the Li, Mg, Al species causing the splitting of the two Ge lone pairs and creating a third ELI–D attractor, and the highly polar bonding Ge–(Li, Mg) as characterized by the corresponding bond fraction $p(\mathrm{lp}^{\mathrm{Li,Mg}})$ being 0.02 and 0.06 for $M = $ Li and Mg (Tables A4.7 and A4.8), respectively. The less polar Al case with $p(\mathrm{lp}^{\mathrm{Al}}) = 0.16$ already represents the transition to the even more covalent Zn case. For $M = $ Zn (and the remaining transition metals) this splitting is no more observed because the polarity of Ge–Zn bonding is even less than the Ge–La one. This can be seen (Tables 4.16 and A4.7–A4.10) from the large increase of covalent characters of the Ge lone pair type regions along $M = $ Li, Mg, Al, Zn for all the (3b)Ge species featuring only 2La + 1 M atoms in the access set.

Continuation of the analysis with transition metals to the left of Zn, namely for $M = $ Cu, Ag, and Pd, is not possible with the same technique, because for the PSC0 correction the active number of valence electrons for the metal is to be specified. For example, for Cu in this compound one can find certain reasons to support either value of 1, or 2, or something in–between. Because of this uncertainty, both possibilities have been exemplarily tested for $M = $ Cu. As a result, with ELIBON(Cu) = + 2 the Ge–Cu polarity slightly decreases with respect to Ge–Zn, while for ELIBON(Cu) = +1 it slightly increases (Tables 4.13 and A4.11). Similar results are also obtained in the case of Y_2PdGe_6, assuming an unlikely Pd^{2+} (Tables 4.14 and A4.14). For La_2AgGe_6 and La_2PdGe_6, the PSC0 correction was not applied, even exemplarily, due to the presence of Ag/Pd penultimate shell bulges (see below) which push this shell into valence region, touching the La penultimate shell basin. As a result, it is not possible to correctly apply the PSC0 correction. Nonetheless, similar results are expected also in the latter cases, also considering obtained values before PSC0 (Tables A4.12–A4.13).

To summarize the investigation of Ge–metal bond polarity, all the title compounds display significant bonding character of the Ge lone–pair type regions due to polar–covalent Ge–La bonding. Along $M = $ Li, Mg, Al, Zn (and Cu, Ag, Pd) the Ge lone pair character is further decreased by Ge–M bonding. Since the lone–pair type basin regions are significantly intersected ("coordinated") by one Ge and three to five neighbouring metal QTAIM atoms (Fig. 4.18), the picture of polyatomic polar bonding evolves. This finding corroborates similar ideas in the literature [27], where such detailed data have not been available.

For $M = $ Li, Mg the small portions of covalent Ge–M bonding can be conceptually omitted, such that the remaining significant covalent Ge–La bonding suggests them to be classified as germanolanthanates $M[La_2Ge_6]$. For the other compounds of this series the metal atoms M have to be included into the covalent framework as well,

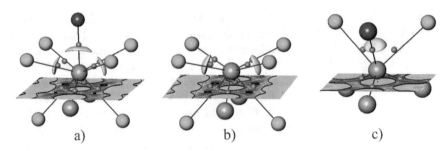

Fig. 4.18 ELI–D isosurfaces (yellow) of Ge lone pair type regions and $-\nabla^2 \widetilde{\Upsilon}_D^\sigma(\mathbf{r})/\widetilde{\Upsilon}_D^\sigma(\mathbf{r})$ attractors (small red spheres) displayed for the coordination set s_C of (2b)Ge2 (**a**), (2b)Ge3 (**b**) and (3b)Ge (**c**) within La_2MgGe_6 unit cell, chosen as a representative. Coloured spheres indicate La (green), Ge (orange) and Mg (brown). Adapted with permission from [36]

which yields a more complex polar intermetallic phase, where even metal–metal interactions have to be taken into account and will be shown below.

Since La_2AgGe_6 is the compound showing the best separation between different DOS regions (Figs. 4.15 and 4.19 (top)), it was decided to calculate the partial ELI–D (pELI–D) in order to check if some Ge–La/Ag interactions could be traced also with this tool.

The lower energy range (4.19a) is mainly dominated by Ge $4s$ states, whereas the region (b) by the more localized Ag $4d$ ones. The interval closer to the Fermi level (c) is mainly dominated by Ge $4p$ and La $5d$ states. The number of electrons per energy region, obtained after integration of both DOS and partial electron density, are in good agreement and sum up to the expected 328 valence electrons per unit cell. It was also checked that the sum of three generated pELI–D yields to the valence ELI–D with the same shape, as the total ELI–D, with tiny lower values.

As expectable, the structuring of Ag penultimate shell is very well visible in the pELI from region (b). Attractors corresponding to (2b)Ge bonds appear only for the pELI from region (a). The same is not true from the (3b)Ge where the same attractors are present both in region (a) and (c). Lone–pairs are again present in both regions, (a) and (c), even though the highest values lays in region (c), closer to the Fermi level as one would expect for states associated to lone–pairs. The main difference, neglecting the pELI values, is their position. In region (a), where there are no La d states, the lone–pairs point directly to Ag atoms. Contrary, in region (c), they are located exactly between the closest La atoms. The position of the corresponding total ELI–D lone–pair attractors is in between, resulting from the sum of the previous including then covalent interaction with both Ag and La. In this exemplary analysis for La_2AgGe_6, it is interesting to highlight that information from DOS/pDOS combined with the derived pELI allows to give insight into chemical bonding.

It is now especially interesting to see, that an effective polycoordination of the Ge lone pair type of basins can be even extracted from the ELI–D distribution alone, without the extra ELI–D/QTAIM basin intersection technique. In a previous study [69], it was discussed, that the regions of the negative ELI–D relative Laplacian

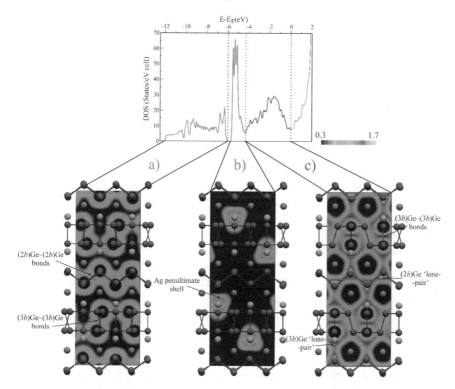

Fig. 4.19 Colour–coded partial ELI–D (pELI–D) (bottom) for the three valence DOS (top) regions **a**, **b** and **c** of La$_2$AgG$_6$. Coloured spheres indicate La (green), Ge (orange) and Ag (grey)

$(-\nabla^2 \tilde{\Upsilon}_D^\sigma(r)/\tilde{\Upsilon}_D^\sigma(r))$ display signatures of bonding interactions on a continuous scale, useful in cases where competing influences prevent the appearance of an ELI–D attractor, *e.g.*, between Fe atoms in Fe$_2$(CO)$_9$ [70]. In the present work, for the first time the ELI–D fine structure was investigated by analysis of the attractor locations of the negative relative ELI–D Laplacian $-\nabla^2 \tilde{\Upsilon}_D^\sigma(r)/\tilde{\Upsilon}_D^\sigma(r)$ distribution. It is to be noted, that the number and location of attractors of $-\nabla^2 \tilde{\Upsilon}_D^\sigma(r)/\tilde{\Upsilon}_D^\sigma(r)$ inside an ELI–D basin is not predetermined by the occurrence of the ELI–D basin attractor. Even if there is only one $-\nabla^2 \tilde{\Upsilon}_D^\sigma(r)/\tilde{\Upsilon}_D^\sigma(r)$ attractor, it need not be at the exactly same position as the ELI-D attractor. It can also happen, that an ELI-D basin does not include a $-\nabla^2 \tilde{\Upsilon}_D^\sigma(r)/\tilde{\Upsilon}_D^\sigma(r)$ attractor, *e.g.* in basins not touching any atomic core region.

For the title compounds it was found, that the total number and the location of $-\nabla^2 \tilde{\Upsilon}_D^\sigma(r)/\tilde{\Upsilon}_D^\sigma(r)$ attractors of a basin $\mathcal{B}$ depicts the effectively coordinating atomic neighbours defining the so-called coordination set $s_C(\mathcal{B})$ of this basin. The interesting observation now is, that for the Ge lone pair type of basins the coordination set s_C is the same as the effective basin atomicity determined from ELI-D/QTAIM basin intersections, *i.e.* the effective intersection set $s_I^{eff}(\mathcal{B})$ of the basin. In detail, each of the lone pair type basins of ELI-D (with one $\widetilde{\Upsilon}_D(r)$ attractor) includes a number of $-\nabla^2\widetilde{\Upsilon}_D^{\sigma}(r)/\widetilde{\Upsilon}_D^{\sigma}(r)$ attractors consistent with the chemically interacting M, La atoms (Fig. 4.18). In contrast, for the Ge–Ge bond basins only these two Ge atoms were found in the coordination set indicated by $-\nabla^2\widetilde{\Upsilon}_D^{\sigma}(r)/\widetilde{\Upsilon}_D^{\sigma}(r)$ attractors. Thus, the locations of the $-\nabla^2\widetilde{\Upsilon}_D^{\sigma}(r)/\widetilde{\Upsilon}_D^{\sigma}(r)$ attractors define a chemically relevant fine structure of the ELI-D distribution not seen before. This feature is mostly a property of the $-\nabla^2\widetilde{\Upsilon}_D^{\sigma}(r)$ distribution and is rather independent of the division by the local ELI-D value. The reason why $-\nabla^2\widetilde{\Upsilon}_D^{\sigma}(r)/\widetilde{\Upsilon}_D^{\sigma}(r)$ has been selected for analysis, is based on plans for future studies, namely its advantageous decomposition into a scaled density Laplacian, a scaled pair-volume function Laplacian and a mixed gradient term [69] (Eq. 3.21). Furthermore, a comparison of the attractor values seems to be mathematically more meaningful for the scaled ELI-D Laplacian.

With this tool at hand, we can now come back to the sudden disappearance of the additional Ge–M ELI–D attractor when changing M from Al to Zn discussed above. As can be seen (Fig. 4.20), the missing ELI–D attractor for $M = $ Zn has converted into a $-\nabla^2\widetilde{\Upsilon}_D^{\sigma}(r)/\widetilde{\Upsilon}_D^{\sigma}(r)$ attractor, and the statement, that Ge–Zn bonding becomes

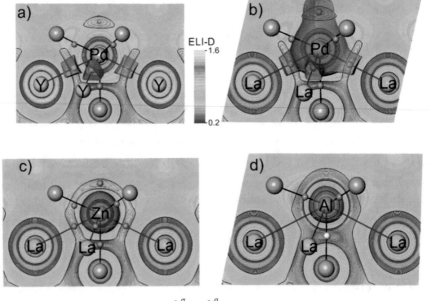

Fig. 4.20 ELI–D distribution with $\nabla^2\widetilde{\Upsilon}_D^{\sigma}(r)/\widetilde{\Upsilon}_D^{\sigma}(r)$ isolines (red for negative values and black for value zero) for Y_2PdGe_6 (**a**), La_2PdGe_6 (**b**) La_2ZnGe_6 (**c**) and La_2AlGe_6 (**d**). Small white (in **a**) and **d**)) and red spheres show attractor positions of ELI-D and $-\nabla^2\widetilde{\Upsilon}_D^{\sigma}(r)/\widetilde{\Upsilon}_D^{\sigma}(r)$, respectively. Reproduced with permission from [36]

similar to Ge–La bonding is corroborated by the three additionally developed Ge–La attractors of $-\nabla^2 \tilde{\Upsilon}_D^{\sigma}(r)/\tilde{\Upsilon}_D^{\sigma}(r)$ (Fig. 4.18).

Moreover, only with the aid of this tool, M–R ($R = Y$, La) bonding interactions in the title compounds can be analysed in a consistent way. Systematic position–space analysis of T–R bonding (T transition metal species) has started with the observation of Mo–4La polyatomic ELF (electron localization function) basin in the carbometallate $La_2[MoC_2]$[71]. In subsequent studies on bi– [72], tri– [73], tetra– [74] and pentametallic [75] organometallic clusters, most of them featuring unsupported T–R bonding signified by an ELI–D attractor close to the internuclear line, the picture of a chemically significant bonding interaction could be established [76]. It has been described as the polar–covalent interaction between an electron–rich transition metal species T acting as the Lewis base, and a Lewis acidic rare earth species R. Transfer of this picture to T–R bonding in intermetallic phases has been performed in parallel [31, 74, 77]. Similarly, polar T–Ba bonding has been previously reported in clathrates $Ba_8T_6Ge_{40}$ [78]. While in all these compounds T–R/Ba bonding were found to be indicated by corresponding ELI–D attractors in the valence region, in the investigated 2:1:6 compounds the situation is no longer so uniform. The largest similarity to those previous works is found for Y_2PdGe_6, where Pd–Y bonding is indicated by four ELI–D attractors between Pd and each of its Y neighbours (Fig. 4.20a). However, already for La_2PdGe_6 these attractors have disappeared. Instead, bulges of the penultimate shell region were observed extending into what would normally be the valence region (Fig. 4.20b). In these bulges now, attractors of $-\nabla^2 \tilde{\Upsilon}_D^{\sigma}(r)/\tilde{\Upsilon}_D^{\sigma}(r)$ could be found, at positions similar to the valence shell attractors in the Pd–Y compound. The same scenario is found for La_2AgGe_6. The valence d orbitals of a transition metal of period n have the main quantum number n–1. Since ELI–D reveals atomic shell structure, the main population of these d orbitals in the free atom are located in the spatial region of the n–1 shell. Depending on the bonding situation, these orbitals can become either more localized or more itinerant, *i.e.* displaying less or more of their electronic population, respectively, in the n shell valence region. Thus, in La_2PdGe_6 the Pd–La interaction seems to display a smaller Lewis basicity–acidity difference, so that the Pd $4d$ orbitals become more localized than in the Pd–Y case. Since $LaPdGe_3$, the Ge–richest compound in the La–Pd–Ge system after La_2PdGe_6, shows a very similar local structure of Pd, an additional chemical bonding analysis was performed. In the latter case, Pd–La bonding (see Fig. 4.21) was found to be indicated by ELI–D attractors in the valence region, as for Pd–Y in Y_2PdGe_6. Thus, it is clear that the scenario described for Pd–La interactions is specific for La_2PdGe_6.

Interestingly, the "uncoordinated" $-\nabla^2 \tilde{\Upsilon}_D^{\sigma}(r)/\tilde{\Upsilon}_D^{\sigma}(r)$ attractor on top of the Pd atoms (Fig. 4.20a and b) in La_2PdGe_6 and Y_2PdGe_6, now has become a coordinated ELI–D Pd–La attractor as well (Fig. 4.21). For La–(Cu, Zn) the situation was found to correspond to the next step of decreasing itinerancy of the penultimate shell orbitals. As shown for $M = Zn$ (Fig. 4.20c) the penultimate shell was rather spherical, and the valence shell does not show ELI–D attractors any more, but only $-\nabla^2 \tilde{\Upsilon}_D^{\sigma}(r)/\tilde{\Upsilon}_D^{\sigma}(r)$ attractors between M–La. Finally, approaching $M = Al$, Mg, Li, even these attractors

Fig. 4.21 ELI–D distribution with $\nabla^2 \widetilde{\Upsilon}_D^{\sigma}(r)/\widetilde{\Upsilon}_D^{\sigma}(r)$ isolines (red for negative values and black for value zero) for LaPdGe₃ intermetallic. White and red spheres are located at the position of ELI–D and $-\nabla^2 \widetilde{\Upsilon}_D^{\sigma}(r)/\widetilde{\Upsilon}_D^{\sigma}(r)$ attractors, respectively

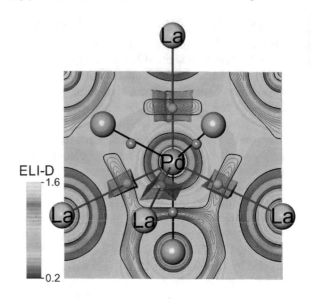

have disappeared, with the only attractor seen in Fig. 4.20d being the ELI–D one for the unusual Ge2 lone–pair type region displaying Ge–MLa₄ polar–covalent bonding discussed above.

So, the series of compounds investigated represents also an extension of the previous T–R bonding studies beyond the $T^{(8)}$ (*i.e.* Fe) group. While in those studies T–R bonding was always revealed by an ELI–D attractor, now a systematic trend towards decrease of the covalent bonding signature was found along Y–Pd, La–{Pd, Ag}, La–{Cu, Zn}, La–{Al, Mg, Li}. For M = Al, Mg, Li no such signature could be observed any more. As a note, in the hypothetical molecule Cp₂Li–YCp₂ (Cp = C₅H₅⁻), studied in [76] to contrast its Li–Y bonding situation with the one in Cp₂Re–YCp₂, besides no ELI–D attractor, also no $-\nabla^2 \widetilde{\Upsilon}_D^{\sigma}(r)/\widetilde{\Upsilon}_D^{\sigma}(r)$ attractor can be found (using the original wave function) either between Li and Y [79] being consistent with the present results for the {Al, Mg, Li} situation. This detailed analysis was only possible with the new ELI–D technique based on the analysis of the $-\nabla^2 \widetilde{\Upsilon}_D^{\sigma}(r)/\widetilde{\Upsilon}_D^{\sigma}(r)$ attractor positions.

Finally, turning to the conceptual classification of the title compounds with respect to the Zintl phases and the 8–N rule, it is clear, that this can be achieved only after careful calibration of the quantum chemical results with respect to reference systems. Basic calibration with respect to ideal behaviour has been achieved previously [33]. The clear deviations from the ideal systems displayed by the title compounds beyond tolerable extent, place them away from the model 8–N scenario. This can be seen from the access electron counts of the Ge species, which are significantly below the approximate 0.3 e range of deviations observed previously (Tables 4.12 and 4.13). A second point is the number of covalent bonding and lone pair type electrons (Table 4.16). For the two–bonded species Ge2 and Ge3 it can be seen (Fig. 4.22), that the corresponding combinations are significantly lower than the ideal (2*b*; 2*lp*) one,

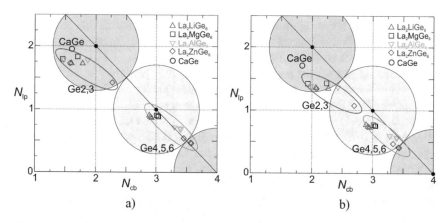

Fig. 4.22 Classification of (2b)Ge species 2, and 3, and (3b)Ge species 4 to 6 in La_2MGe_6 ($M =$ Li, Mg, Al, Zn) and CaGe based on amounts of two–electron covalent bonds (N_{cb}) and lone pairs (N_{lp}). The large grey spheres mark the domains of the (2b, 2lp), (3b,1lp), and (4b, 0lp) scenarios, respectively; **(a)** without PSC0 and **(b)** with PSC0. Picture b) is reproduced with permission from [36]

mainly after the PSC0 correction where the covalent interactions between (2b)Ge lone–pairs and La species have been disclosed (Fig. 4.22b).

This increased covalency degree obtained after PSC0 correction push them close to the boundary of the corresponding (2 b; 2 lp) domain (grey sphere), or in case of species Ge2 for La_2AlGe_6 and La_2ZnGe_6 even within the (3 b;1 lp) domain. Note, that species Ge2 is the one with the highest heteroatomic coordination number, (Figs. 4.11 and 4.18a, La_6 trigonal prism with Mg atom capping one of its rectangular faces). Focusing on the three–bonded Ge species Ge4, 5, and 6, differences without and after PSC0 are less pronounced. In fact, in both cases, they are mainly located within the (3b, 1lp) domain, with the ones for La_2AlGe_6 nearly touching the boundary and La_2ZnGe_6 surpassing the boundary toward the (4b; 0lp) domain. Nevertheless, all the values after PSC0 are pushed at lower N_{lp} values, withdrawing, in particular for La_2LiGe_6 and La_2MgGe_6, from the classical (3b, 1lp) point.

In such cases, where deviations from calibration points are found in all directions of parameter space, it is useful to find additional calibration points, where at least one of the parameters behaves close to ideal. For the present systems the CaGe phase (oS8 structure [44]) was chosen as such a new calibration point.

The CaGe structure contains one–dimensional Ge zigzag chains consistent with the formal $(2b)Ge^{2-}$ picture. It is isotypic to CaSi, for which Ca–Si partially covalent bonds were identified on the basis of the electron density (QTAIM method) and discussed to lead to a significant deviation from the Zintl picture [80]. In this respect it is not unimportant to know, that both CaSi and CaGe display metallic character with only a pseudo gap found in the band structure at the Fermi level. Using the same position–space techniques including PSC0 correction like for the title compounds yields $N_{acc}^{ELI}(Ge) = 7.13$ e, $Q^{eff}(Ge) = -1.26$, $N_{val}(Ge) = 5.26$ e, and a Ge(1.11 be; 0.73 be', 3.41 lpe') scenario. With $N_{acc}^{ELI}(Ge) = 7.13$ e it is clearly out of the tolerable range of (8.0 ± 0.3) e, roughly similar to species Ge2 and 3 for the title compounds. Moreover, there is a decreased effective charge of only -1.26, accompanied by a reduced number of electrons inside the Ge–Ge bond basins (1.11 e instead of 2.0 e). Considering just the net population of the two lone pair type regions 2×2.44 e, a severe overpopulation with respect to the ideal model 2×2.00 e is observed. But after the QTAIM/ELI–D basin intersection analysis, the Ge atoms are found to possess $0.73 + 3.41 = 4.14$ electrons of these total populations, symbolically written as 4.14 pe = 0.73 be' + 3.41 lpe'. The remaining 0.73 electrons of the lone–pair type regions represent the covalent counterparts inside the surrounding 4 Ca atoms yielding a Ge–Ca$_4$ multiatomic polar bonding picture. The 4.14 e contributed by Ge virtually are the equivalent of 2×2 e lone pairs. In total this leads to a partial compensation of the loss of number of covalent electrons for Ge–Ge bonding by polar Ge–Ca bonding. Summing up (be + be') contributions leads 1.11 e + 0.73 e = 1.84 e covalent bonding electrons per Ge. The transformation of this (be; lpe) notation for electron counting to two–electron bonds and lone pair counting (b; lp) is achieved according to b = be, and lp = lpe/2. As a result, with these $(N_{cbe}; N_{lpe})$ characteristics, (1.84 be; 3.41 lpe) = (1.84 b; 1.71 lp), the Ge species of CaGe is located well inside the (2 b; 2 lp) bonding domain for $(2b)Ge^{2-}$ (Fig. 4.22). Its location below the diagonal black line connecting the conceptual (2 b; 2 lp) and (3 b; 1 lp) point indicates a corresponding lack of electrons caused by the incomplete charge transfer and corresponding polar Ge–Ca bonding. This feature for CaGe is used as a new reference point for the calibration.

As a common feature, CaGe and the title compounds reveal a significant lowering of the Ge access electron numbers with respect to 8 electrons. For the latter compounds the largest effect occurs for the $(2b)$Ge species, where specifically $(2b)$Ge3 displays the lowest access electron numbers. Moreover, it can be seen (Fig. 4.22), that the two–bonded Ge species of title compounds clearly deviate from the CaGe situation, with Ge2 for $M = $ Zn showing the largest deviation, and $M = $ Li, Mg being closest to it. The other La_2MGe_6 compounds with $M = $ Cu, Ag, Pd and Y_2PdGe_6 are expected to fall, on the basis of the obtained results without and after the "inappropriate" PSC0 correction (Tables A4.11–A4.14), into the same classification as the Zn one.

4.7 Conclusions on the R_2MGe_6 Family of Ternary Germanides[7]

The R_2PdGe_6 series of compounds was targeted with the aim to elucidate more on the formation, crystal structure and chemical bonding of the R_2MGe_6 family (R = rare earth metal, M = another metal), joining experimental results with structure analysis and quantum chemical calculations. This study was motivated by the controversial and sometimes erroneous structural data available for the title compounds, possibly

---

[7]Part of this section was originally published in [35,36].

also affecting the interpretation of the numerous magnetic measurements performed by several authors and by their intriguing structural motifs suggesting the presence of interesting chemical bonding scenarios.

The R_2PdGe_6 phase was detected in all samples prepared by direct synthesis, except for R = Sc and Eu.

Single crystal X–ray analyses conducted on several representatives (R = Y, Ce, Pr, Nd, Er, Lu) indicated that the $oS72$–$Ce_2(Ga_{0.1}Ge_{0.9})_7$ is the correct structural model, as previously reported for the Dy and Yb analogues. The same crystal structure is suggested for R = Sm, Gd, Tb, Ho, Tm on the basis of powder X–ray diffraction measurements and behavior regularities along the rare–earth series.

Different annealing treatments along with thermal analysis investigations were performed with the aim to obtain good crystals or a sufficiently high yield of La_2PdGe_6; the obtained results suggest that this is a metastable phase likely to form in small amount during slow cooling treatments.

Therefore, the flux synthesis, able to stabilize metastable phases, was explored, using In as a metal solvent. The good quality La_2PdGe_6 crystals obtained after optimizing this method, are non–merohedrally twins of the $mS36$–La_2AlGe_6 structure. Same morphology monoclinic twins of two equally big domains were obtained for the flux–synthesized Pr_2PdGe_6 compound. This result suggests that also La_2PdGe_6 obtained from direct synthesis might be of $oS72$ orthorhombic structure. Instead, for the heavy rare earth representative Yb_2PdGe_6 the flux does not stabilize the monoclinic structure, and the $oS72$ model remains the correct one.

Taking into account the considerable amount of data on 2:1:6 germanides, some structure–rationalizing idea was pursued. Supposing the vacancy ordering phenomenon to be the key–factor of structural changes, a compact Bärnighausen tree was constructed, with rigorous group–subgroup relations between the $oS20$–$SmNiGe_3$ aristotype and the three possible derivatives $oS72$, $mS36$ and $oS18$: their structural models are localized on separate branches of the tree, representing different symmetry reduction paths. The presence of a t2 reduction step on the path bringing to the $mS36$ model is coherent with the formation of twinned crystals.

Both from our experience and literature data [81, 82], binary and ternary germanides are particularly prone to geminate, and this possibility should be carefully investigated during structural solution. For this reason, these phases are also suitable for studies targeted to better understand origin, formation conditions and types of twinning in intermetallics. At present, we are further studying twinned crystals of germanides, where twinning seems to be related to vacancy ordering phenomena.

The same phenomena can lead to modulated structures, as already found for example both for binary RGe_{2-x} [83], representing another reason of interest of the studied compounds.

From the presented symmetry reduction scheme it became obvious the symmetry discrepancy between the $oS18$ (2nd order derivative of the aristotype) and both the $oS72$ and $mS36$ models (4th order derivatives), highlighting the poor crystal chemical reliability of the structure mostly reported in the literature and already corrected for some R_2MGe_6 compounds.

Aiming to confirm this idea and to discard energetically less probable modifications, DFT total energy calculations were performed for R_2PdGe_6 (R = Y, La) in the three abovementioned structural models, and for La_2MGe_6 (M = Li, Pt, Cu, Ag, Au) in the $oS18$ and $oS72$ modifications. Structure optimization was performed before all calculations. Considering the total energy highest values as well as the crystallochemical factors, the $oS18$–Ce_2CuGe_6 ($oS18$–Pr_2LiGe_6 for M = Li) model is the less probable to occur. With the purpose to confirm this idea by experimental results, La_2MGe_6 (M = Cu, Ag) and R_2LiGe_6 (R = La–Nd) where synthesized and structurally characterized. Results confirmed the theoretical expectations; in fact, La_2MGe_6 (M = Cu, Ag) crystallize with $oS72$ structure whereas La_2LiGe_6 with the monoclinic $mS36$ modification (on the basis of powder data, the same structure was suggested for Ce and Pr–containing alloys). It was also concluded that the $oS72$ and $mS36$ models, containing geometrically equivalent fragments, are energetically equivalent polytypes. In fact, they are sensitive to the crystallization conditions, as became clear from results of the conducted flux synthesis. This approach turned out to be an effective alternative method to prepare these compounds.

All the obtained results, which constitute a complete structural revision for the R–M–Ge compounds of ~ 2:1:6 stoichiometry, are represented in Table 4.17.

Table 4.17 Structure types revision of R–M–Ge compounds of ~ 2:1:6 stoichiometry (R = rare earth metal; M = another metal) on the basis of the experimental (red), theoretical (yellow) and both (orange) results

Group	M	R= Sc	Y	La	Ce	Pr	Nd	Sm	Eu	Gd	Tb	Dy	Ho	Er	Tm	Yb	Lu
I	Li			◊	◊	◊											
II	Mg		—	□	—	—	—	—	—	—	—	—	—	—	—	—	—
VII	Mn			⊠	⊠												
IX	Co		⊠							⊠	⊠	⊠	⊠	⊠	⊠		⊠
	Rh						⊠										
X	Ni		⊠	⊠	⊠	⊠	—			⊠	⊠	⊠	⊠	⊠	⊠	⊠	⊠◊
	Pd	—	□	◊	□	□◊	□	□	—	□	□	□	□	□	□	□	□
	Pt		⊠	□◊	⊠	⊠	⊠	⊠	⊠	⊠	⊠	⊠	⊠	⊠	⊠	⊠	
XI	Cu		⊠	□	⊠	⊠	⊠	⊠	—	⊠	⊠	⊠	⊠	⊠	⊠	⊠◊	
	Ag			□	⊠	⊠	⊠	⊠	—	⊠			—				—
	Au			□◊	⊠	⊠	⊠	⊠	—	⊠	⊠	⊠	—				
XII	Zn	◆	□	□	□	□	□			□	□	◆	◆	—	—		
XIII	Al	—	◊	◊	◊	◊	◊		—	◊	◊	⊠	—	—	—	—	—
	Ga		□◊	□			—										

⊠ = $oS18$–Ce_2CuGe_6 SG: *Amm2* (№ 38)   □ = $oS72$–$Ce_2(Ga_{0.1}Ge_{0.9})_7$ SG: *Cmce* (№ 64)

◊ = $mS36$–La_2AlGe_6 SG: *C2/m* (№ 12)   — = not observed

◆ = $mP34$-2–$Dy_2Zn_{1-x}Ge_6$ SG: *P2/m* (№ 11)

It is important to highlight that energy calculations do not allow to distinguish among $oS72$ and $mS36$. That is why they are both reported for La_2PtGe_6 and La_2AuGe_6

The study presented above represents a good and essential basis for investigations concerning the chemical bonding for $R_2M\mathrm{Ge}_6$, which have been conducted applying the recent position–space quantum chemical techniques [36]. In particular, a comparative chemical bonding analysis was performed for the $\mathrm{La}_2M\mathrm{Ge}_6$ ($M = \mathrm{Li}$, Mg, Al, Zn, Cu, Ag, Pd) and $\mathrm{Y}_2\mathrm{PdGe}_6$ germanides by means of QTAIM, ELI–D, and their basin intersections. The presence of zigzag chains and corrugated layers of covalently bonded Ge atoms was confirmed for all compounds. In addition, a detailed study of polar covalent bonding between the germanium polyanion and the metal partial structure was presented. In order to correctly face this non trivial task, the new penultimate shell correction (PSC0) for the electronic occupation of the valence basins was introduced and applied for $M = \mathrm{Li}$, Mg, Al, Zn and also to the binary CaGe, chosen as reference system. In fact, in its current form, this correction is only correctly applicable for metal atoms with an unambiguous oxidation state, which represents a certain restriction to be overcome in future work. Compared to CaGe, which itself is known to deviate from the ideal $8-N$ picture, the deviations found for $\mathrm{La}_2M\mathrm{Ge}_6$ are much more pronounced and indicate a clear trend increasing moving from Mg to Li, Al and Zn. The low tendency of La atoms to complete electron transfer $\mathrm{La} \rightarrow \mathrm{Ge}$ in all investigated compounds, accompanied by an increasing tendency of $M = \mathrm{Li}$, Mg, Al, Zn for similar behaviour, causes polar covalent bonding Ge–($n\mathrm{La}$, M) of multiatomic nature replacing the classically nonbonding Ge lone–pair situation. The particularly small portion of covalent Ge–Li and Ge–Mg bonding allows to conceptually omit it, leading to the description of the Li– and Mg–containing phases as germanolanthanates, with $M[\mathrm{La}_2\mathrm{Ge}_6]$ ($M = \mathrm{Li}$, Mg) formula. The multiatomic character of the Ge–($n\mathrm{La}$, M) bonds was shown to be indicated also by attractors of $-\nabla^2 \widetilde{\Upsilon}_D^{\sigma}(r)/\widetilde{\Upsilon}_D^{\sigma}(r)$, which represents a new proposal for ELI–D fine structure analysis. With increasing number of active valence electrons left on M atoms also interactions in the metallic partial structure gradually develop. With the aid of the ELI–D fine structure analysis tool $-\nabla^2 \widetilde{\Upsilon}_D^{\sigma}(r)/\widetilde{\Upsilon}_D^{\sigma}(r)$ increasing M–R bonding was discussed along $M = \mathrm{La}$–{Li, Mg, Al}, La–{Zn, Cu}, La–{Ag, Pd}, and Y–Pd extending known T–R bonding scenarios in position space presented in previous studies.

Acknowledgements The author thanks *Wiley-VCH* and *Royal Society of Chemistry* for allowing the reuse of its own published works in this thesis.

References

1. Villars P, Cenzual K (2018/19) Pearson's crystal data. ASM International, Ohio, USA
2. Sologub OL, Hiebl K, Rogl P, Bodak OI (1995) J. Alloys Compd. 227:37
3. Konyk M, Romaka L, Gignoux D, Fruchart D, Bodaka O, Gorelenko Yu (2005) J. Alloys Compd. 398:8
4. Shigetoh K, Hirata D, Avila MA, Takabatake T (2005) J. Alloys Compd. 403:15–18
5. Kaczorowski D, Konyk M, Szytula A, Romaka L, Bodak O (1891) Solid State Sci. 10

6. Wawryk R, Tróc R, Gribanov AV (2012) J. Alloys Compd. 520:255–261
7. Kaczorowski D, Konyk M, Romaka L (2012) J. Alloys Compd. 526:22
8. Kaczorowski D, Gribanov AV, Rogl P, Dunaev SF (2016) J. Alloys Compd. 685:957–961
9. Konyk MB, Romaka LP, Gorelenko YuK, Bodak OI (2000) J. Alloys Compd. 311:120
10. Pavlyuk VV, Pecharskii VK, Bodak OI (1988) Kristallografiya 33:43
11. Stetskiv A, Misztal R, Pavlyuk VV (2012) Acta Crystallogr. C 68:i60
12. Pavlyuk VV, Bodak IO (1992) Dopov. Akad. Nauk Ukr. 2:78–79
13. Rizzoli C, Sologub O, Salamakha P (2003) J. Alloys Compd. 351:L10
14. Solokha P, De Negri S, Proserpio DM, Blatov VA, Saccone A (2015) Inorg. Chem. 54:2411. https://doi.org/10.1021/ic5030313
15. Gribanov A, Safronov S, Murashova E, Seropegin Y (2012) J. Alloys Compds 542:28
16. Fornasini ML, Manfrinetti P, Palenzona A (2002) Z. Kristallogr. 217*NCS,* 173
17. De Negri S, Solokha P, Skrobańska M, Proserpio DM, Saccone A (2014) J. Solid State Chem. 218:184–195. https://doi.org/10.1016/j.jssc.2014.06.036
18. Peter SC, Subbarao U, Sarkar S, Vaitheeswaran G, Svane A, Kanatzidis MG (2014) J. Alloys Compds 589:405
19. Tokaychuk YO, Filinchuk YE, Fedorchuk AO, Kozlov AYu, Mokra IR (2006) J. Sol. State Chem. 179:1323–1329
20. Grin Y (1992) In: Parthè E (ed) Modern perspectives in inorganic crystal chemistry. Kluwer Academic Publishers, Dordrecht, the Netherlands, 77
21. Zhao JT, Cenzual K, Parthè E (1991) Acta Cryst. C 47:1777–1781
22. Müller U (2006) Inorganic structural chemistry, 2 edn. Wiley
23. Klemm W, Busmann E, Anorg Z (1963) Allg. Chem. 319:297–305
24. Mudring AV, Corbett J (2004) J. Am. Chem. Soc. 126:5277–5281
25. Ponou S, Doverbratt I, Lidin S, Miller GJ (201) Eur J Inorg Chem 169–176.
26. Miller GJ, Schmidt MW, Wang F, You T-S (2011) Quantitative advances in the Zintl–Klemm formalism. In: Fässler TF (ed) Zintl phases, principles and recent developments. Springer Series in Structure and Bonding 139, Springer Science+Business Media Dordrecht
27. Ponou S, Lidin S (2016) Dalton Trans. 45:18522–18531
28. Bader RFW (1990) Atoms in molecules: a quantum theory. Oxford University Press, Oxford
29. Wagner FR, Baranov AI, Grin Yu, Kohout M, Anorg Z (2013) Allg. Chem. 639(11):2025–2035
30. Börrnert C, Grin Yu, Wagner FR, Anorg Z (2013) Allg. Chem. 639(11):2013–2024
31. Dashjav E, Prots Yu, Kreiner G, Schnelle W, Wagner FR, Kniep R (2008) J. Solid State Chem. 181:3121–3130
32. Baitinger M, Böhme B, Ormeci A, Grin Y (2014) Chapter 2 solid state chemistry of clathrate phases: crystal structure. Chemical bonding and preparation routes. In: Nolas GS (ed) The physics and chemistry of inorganic clathrates. Springer Series in Materials Science 199, Springer Science+Business Media Dordrecht
33. Bende D, Wagner FR, Grin Yu (2015) Inorg. Chem. 54:3970–3978
34. Wagner FR, Bende D, Grin Yu (2016) Dalton Trans. 45:3236–3243
35. Freccero R, Solokha P, Proserpio DM, Saccone A, De Negri S (2017) Dalton Trans. 46:14021–14033
36. Freccero R, Solokha P, De Negri S, Saccone A, Grin Yu, Wagner FR (2019) Polarcovalent bonding beyond the Zintl picture in intermetallic rare earth germanides. Chem. Eur. J. 25:6600–6612. https://doi.org/10.1002/chem.201900510CopyrightWiley-VCHVerlag GmbH&Co.KGaA.Reproducedwithpermission
37. Giannozzi P, Baroni S, Bonini N, Calandra M, Car R, Cavazzoni C, Ceresoli D, Chiarotti GL, Cococcioni M, Dabo I, Dal Corso A, de Gironcoli S, Fabris S, Fratesi G, Gebauer R, Gerstmann U, Gougoussis C, Kokalj A, Lazzeri M, Martin-Samos L, Marzari N, Mauri F, Mazzarello R, Paolini S, Pasquarello A, Paulatto L, Sbraccia C, Scandolo S, Sclauzero G, Seitsonen AP, Smogunov A, Umari P, Wentzcovitch RM (2009) J. Phys. Condens. Matter 21:395502
38. Perdew JP, Burke K, Ernzerhof M (1996) Phys. Rev. Lett. 77:3865
39. Vanderbilt D (1990) Phys. Rev. B 41:7892
40. GBRV high-throughput pseudopotential, https://physics.rutgers.edu/gbrv. Accessed May 2017

41. Blum V, Gehrke R, Hanke F, Havu P, Havu V, Ren X, Reuter K, Scheffler M (2009) Comput. Phys. Commun. 180:2175–2196
42. Program FHI-aims, an all electron, numerically tabulated, atom-centered-orbital electronic structure code, version 071914_7; Fritz-Haber-Institut, Berlin, Germany.
43. Troc R, Wawryk R, Gribanov AV (2013) J. Alloys Compd. 581:659–664
44. Iandelli A, Rolla SL (1955) Atti Accad. Naz. Lincei, Cl. Sci. Fis., Mat. Nat., Rend. 19:307–313
45. Koepernik K, Eschrig H (1999) Phys. Rev. B 59:1743
46. Perdew JP, Wang Y (1992) Phys. Rev., B 45:13244–13249
47. Massalski TB (1990) Binary alloy phase diagrams, vol. 2. American Society for Metals, Metals Park, Oh 44073, USA
48. Kanatzidis MG, Pöttgen R, Jeitschko W (2005) Angew. Chem. Int. Ed. 44:6996
49. Solokha P, Freccero R, De Negri S, Proserpio DM, Saccone A (2016) Struct. Chem. 27:1693–1701
50. Salvador JR, Gour JR, Bilc D, Mahanti SD, Kanatzidis MG (2004) Inorg. Chem. 43:1403
51. Konyk MB, Salamakha PS, Bodak OI, Pecharskii VK (1988) Sov. Phys. Crystallogr. 33:494–495
52. Bardin OL, Bodak OI, Belan BD, Kryvulya LV, Protsyk OS (1999) Visn. Lviv. Derzh. Univ., Ser. Khim 38:58–63
53. Petricek V, Dusek M, Palatinus L (2014) Z. Kristallogr. 229(5):345–352
54. Sheldrick GM (2008) Acta Cryst. A 64:112–122
55. Gelato LM, Parthé E (1987) J. Appl. Crystallogr. 20:139
56. Müller P (2006) Crystal structure refinement: a crystallographer's guide to SHELXL. Oxford University Press, Oxford, UK
57. Clegg W (2009) Crystal structure analysis: principle and practice, 2nd edn. Oxford University Press, Oxford, UK
58. Farrugia LJ (1999) J. Appl. Cryst. 32:837–838
59. Bruker (2014) APEX2, SAINT-Plus, XPREP, SADABS, CELL_NOW and TWINABS. Bruker AXS Inc., Madison, Wisconsin, USA
60. Shannon RD (1976) Acta Cryst. 32:751
61. Peter SC, Chondroudi M, Malliakas CD, Balasubramanian M, Kanatdidis MG (2011) J. Am. Chem. Soc. 133:13840–13843
62. Christensen J, Lidin S, Malaman B, Venturini G (2008) Acta Cryst. B 64:272–280
63. Müller U (2013) Symmetry relationships between crystal structures. applications of crystallographic group theory in crystal chemistry. Oxford University Press, Oxford, UK
64. Prince E (2004) International tables for crystallography, vol C, Mathematical, physical and chemical tables. Kluwer Academic Publisher, Dordrecht
65. Wagner FR, Bezugly V, Kohout M, Grin Yu (2007) Chem. Eur. J. 13:5724–5741
66. Bronger W, Baranov AI, Wagner FR, Kniep R, Anorg Z (2007) Allg. Chem. 633:2553–2557
67. Kohout M, Savin A (1996) Int. J. Quantum. Chem. 60:875–882
68. Baranov AI (2014) J. Comput. Chem. 35:565–585
69. Wagner FR, Kohout M, Grin Yu (2008) J. Phys. Chem. 112:9814–9828
70. Wagner FR, Cardoso-Gil R, Boucher B, Wagner-Reetz M, Sichelschmidt J, Gille P, Baenitz M, Grin Yu (2018) Inorg. Chem. 57:12908–12919
71. Dashjav E, Kreiner G, Schnelle W, Wagner FR, Kniep R, Anorg Z (2004) Allg. Chem. 630:689–696
72. Butovskii MV, Tok OL, Wagner FR, Kempe R (2008) Angew. Chem. Int. Ed. 47:6469–6472
73. Döring C, Dietel A-M, Butovskii M, Bezugly V, Wagner FR, Kempe R (2010) Chem. Eur. J. 16:10679–10683
74. Butovskii MV, Döring C, Bezugly V, Wagner FR, Grin Yu, Kempe R (2010) Nat. Chem. 2:741–744
75. Butovskii M, Tok OL, Bezugly V, Wagner FR, Kempe R (2011) Angew. Chem. Int. Ed. 50:7695–7698
76. Butovskii MV, Oelkers B, Bauer T, Bakker JM, Bezugly V, Wagner FR, Kempe R (2014) Chem. Eur. J. 20:2804–2811

77. Davaasuren B, Borrmann H, Dashjav E, Kreiner G, Widom M, Schnelle W, Wagner FR, Kniep R (2010) Angew. Chem. Int. Ed. 49:5688–6592
78. Ormeci A, Grin Yu (2015) J. Thermoelec. 6:16–32
79. Wagner FR unpublished results
80. Kurylyshyn IM, Fässler TF, Fischer A, Hauf C, Eickerling G, Presnitz M, Scherer W (2014) Angew. Chem. Int. Ed. 53:3029–3032
81. Budnyk S, Weitzer F, Kubata C, Prots Y, Akselrud LG, Schnelle W, Hiebl K, Nesper R, Wagner FR, Grin Y (2006) J. Solid State Chem. 179:2329
82. Shcherban O, Savysyuk I, Semuso N, Gladyshevskii R, Cenzual K (2009) Chem. Alloy. Met. 2:115
83. Tobash PH, Lins D, Bobev S (2006) Inorg. Chem. 45:7286

Chapter 5
On the Existence, Crystal Structure and Physical Properties of $R_2Pd_3Ge_5$ (R = Rare Earth Metal) Intermetallic Germanides

Ternary R–T–Ge systems (R = rare earth metal; T = transition element) are very rich in intermediate compounds, which have been extensively studied from the points of view of synthesis, crystal structure and physical properties [1–5]. During the doctorate research work, an accurate revision (see this chapter) of the R_2PdGe_6 series (R = Y, La–Nd, Sm, Gd–Lu), complemented by new syntheses, crystal structure determinations and energy calculations, revealed the existence of two 2:1:6 polytypes, whose stability is sensitive to the crystallization conditions, as follows from results of flux syntheses [6]. That is why, among the R–T–Ge systems, the attention was focused on the R–Pd–Ge compounds where more than one hundred ternary phases are known. Their compositional variety is plotted in Fig. 5.1.

Most of these compounds are stoichiometric and distributed over the whole compositional range, with the exception of the R rich corner. Noticeable differences can be highlighted between light (L) and heavy (H) rare earth metals. Pd-rich representatives were discovered only in combination with light rare earths, whereas more extensive studies were conducted for Ge-richer compounds (region evidenced in Fig. 5.1). In this framework, one may note several families of compounds existing for almost all rare earths, for example RPd_2Ge_2 [8] and R_2PdGe_6 [6]; other series, such as $RPdGe$ [9, 10] and $R_3Pd_4Ge_4$ [9], show a structural change with the temperature or as a function of R atomic number. Some compounds were reported only with light ($R_2Pd_3Ge_5$ [11–15], $RPdGe_3$ [16, 17]) or heavy ($RPdGe_2$ [18], $R_5Pd_4Ge_8$ [19]) rare earth metals. Along the 33.3 at.% R isoconcentration line two families of compounds were reported (AlB_2-like and $CeNiSi_2$-like) the compositions of which slightly change depending on R. In the former case the Pd and Ge atoms share the same crystallographic site and thus the $R(Pd_xGe_{1-x})_2$ formula is the most appropriate; instead, in the latter case the Pd site is partially occupied leading to the RPd_xGe_2 formula.

© The Editor(s) (if applicable) and The Author(s), under exclusive license
to Springer Nature Switzerland AG 2020
R. Freccero, *Study of New Ternary Rare-Earth Intermetallic Germanides
with Polar Covalent Bonding*, Springer Theses,
https://doi.org/10.1007/978-3-030-58992-9_5

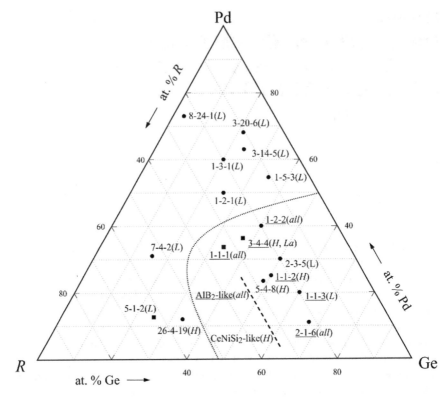

Fig. 5.1 Known *R*–Pd–Ge ternary compounds plotted on Gibbs triangle. Stoichiometries indicated with ■ correspond to more than one structural modification; (*L*) and (*H*) stay for light and heavy rare earth elements, respectively; (all) indicates almost complete series of compounds along *R*. The dashed line includes compositions for which the AlB₂ and CeNiSi₂ type structures were reported. Yb-containing compounds are underlined. The dotted line delimits ternary compounds of major interest for this work [7]

During the investigations (described in Chap. 4) on the R_2PdGe_6 series the new $Nd_2Pd_3Ge_5$ was serendipitously found and characterized by single crystal X-ray diffraction data. This finding encouraged a more accurate literature data search on the existence and crystal structure of $R_2T_3Ge_5$ phases. Two closely related structural modifications (*oI*40 and *mS*40) were found and their distribution is represented in Table 5.1.

Another crystal structure (SG: *Pmmn*) was proposed for some compounds of this family, but no clear structural model was presented [22]. The 2:3:5 stoichiometry was observed for all the 9th and 10th group transition elements, in some cases extending along the lanthanide series and including Yttrium. A structural change from the *oI*40 to the *mS*40 modification was found for $T = \text{Co, Rh}$ on increasing the R atomic number. For the two compounds $Pr_2Co_3Ge_5$ and $Nd_2Co_3 \, Ge_5$ both structure-types were detected: the orthorhombic one in samples annealed at 870 K [24, 25] and the monoclinic one in samples annealed at 1173 K [4].

Table 5.1 Distribution of structure types of $R_2T_3Ge_5$ compounds (R = rare earth metal; T = 9th or 10th group metal). Reproduced with permission from [20]

R / T	Y	La	Ce	Pr	Nd	Sm	Gd	Tb	Dy	Ho	Er	Tm	Yb	Lu	Refs.
Co	■	□	□	■ □	■ □	■	■	■	■	■	■				[1, 21]
Rh	■	□	□	□	■	■	■	■	■	■	■	■			[1]
Ir	□	□	□	□	□	□	□	□	□	▦	▦	▦	▦	■	[1, 22]
Ni			□	□											[1]
Pd		□	□	□		□									[1, 11–13, 23]
Pt		□		□		□									[1]

□ = $oI40$-U$_2$Co$_3$Si$_5$; ■ = $mS40$-Lu$_2$Co$_3$Si$_5$; ▦ = not completely determined, SG: *Pmmn*

On the other hand, less representatives were revealed for the 10th group transition elements; for $T = $ Pd and $R = $ La, Ce, Pr, Sm the $oI40$-$U_2Co_3Si_5$ crystal structure was only determined by X-ray powder diffraction data [11–13, 23]. The new results, presented below, for the Nd analogue, complete the series with the light rare earth-containing compounds. Anyway, as described in the first part of this chapter, samples with $R = $ La, Ce, Pr, Sm were synthesized in order to check their structures and existence. Whereupon, the existence of $R_2Pd_3Ge_5$ compounds along the whole lanthanide series (including then alloys with $R = $ Gd–Lu), was performed. In both cases (all R), explorative syntheses were conducted on selected $R_2Pd_3Ge_5$ by using different metal fluxes, such as In, Bi and Pb and the preliminary obtained results, including also the separation techniques, are reported in the paragraphs below.

In this work, the crystal structure of the title germanides is also presented applying the concept of symmetry reduction as derivative from the RPd_2Ge_2, belonging to the huge family of the derivatives of the ubiquitous $BaAl_4$ aristotype.

Moreover, since compounds belonging to the $R_2T_3X_5$ group ($T = $ transition element, $X = $ Ga, Si, Ge, Sn) exhibit interesting physical properties (*e.g.* superconductivity, different magnetic transitions, heavy fermion behavior, etc....) [26–28], the new $Nd_2Pd_3Ge_5$ and $Yb_2Pd_3Ge_5$ compounds were selected for such properties measurements. In this Chapter, results for the Yb-containing intermetallic are reported whereas those for the Nd one are still under investigation.

5.1 The $R_2Pd_3Ge_5$ Intermetallics with Light Rare Earth Metals (La-Nd, Sm): Synthesis, Crystal Structures and Structural Relations[1]

5.1.1 Sample Synthesis

Different synthetic routes were followed in this work, including direct synthesis in induction, resistance and arc-furnace and flux synthesis. The compounds $Nd_2Pd_3Ge_5$ was detected in one sample (total mass ~0.8 g) of nominal composition $Nd_{22.2}Pd_{11.1}Ge_{66.7}$, synthesized in a resistance furnace. The stoichiometric amounts of the pure elements were enclosed in an arc-sealed Ta crucible, which was then closed in an evacuated quartz phial to prevent oxidation at high temperature, and finally placed in a resistance furnace, where the following thermal cycle was applied: 25 °C → (10 °C/min) → 950 °C (1 h) → (−0.2°C/min) → 350 °C → (furnace switched off) → 25 °C. A continuous rotation, at a speed of 100 rpm, was applied to the phial during the thermal cycle.

Analogue compounds with $R = $ La, Ce, Pr and Sm were characterized in as-cast samples with nominal composition $R_{20.0}Pd_{30.0}Ge_{50.0}$, prepared by arc- (La, Ce, Pr) or induction (Sm) melting, with the purpose to confirm the crystal structure. The

---

[1]Part of this section was originally published in [20].

induction melting procedure was preferred for the Sm-containing sample, in order to avoid any evaporation of this element, which is characterized by a quite high vapour pressure. For this sample, the weighed constituent elements were enclosed in an arc-sealed tantalum crucible and induction melted. The melting was performed under a stream of pure argon and it was repeated several times in order to ensure homogeneity. The Ce and Pr containing samples were synthesized by repeated arc melting the elements on a water-cooled copper hearth with a tungsten electrode. Weight losses after arc-melting were generally less than 1%. In order to obtain a sample suitable for physical properties measurement, an alloy with nominal composition $Nd_{20.0}Pd_{30.0}Ge_{50.0}$ was also prepared by arc melting, according to the just mentioned procedure.

In order to investigate the influence of metal flux on phase formation, syntheses with metal solvents were also performed. Lanthanum and Neodymium were chosen for this purpose. Hence, the stoichiometric amounts of elements were put in an arc-sealed Ta crucible with a 1:4 molar excess of metal flux (Bi for the La containing sample; In, Pb and Bi for the Nd containing sample). The total mass of each sample was about 4 g. Then the Ta crucible was closed in an evacuated quartz phial and placed in a resistance furnace, applying the following thermal cycle: 25 °C → (8 °C/min) → 1050 °C (2 h) → (−8.0 °C/min) → 300 °C → (furnace switched off) → 25 °C. A continuous rotation of the quartz phial during the thermal cycle was applied to favour a better dissolution of the constituting elements inside the flux. A vertical cut of the obtained ingots revealed the presence of large shining crystals, visible to the naked eye, randomly distributed within the flux-solidified matrix. For the sample synthesized with Pb flux, a separation of the solvent through centrifugation at T > $T_m(Pb)$ was tested.

5.1.2 Samples Characterization by SEM/EDXS Analysis

Samples synthesized by slow cooling in the resistance furnace are characterized, as usual, by an inhomogeneous microstructure, containing several different phases, concentrated in the form of large crystals. In the Nd-containing sample for which the nominal composition is farther from the 2:3:5 stoichiometry, the $Nd_2Pd_3Ge_5$ compound coexists with Nd_2PdGe_6, which is the principal phase, and with other unknown ternary phases (see Table 4.2).

All samples prepared by arc- or induction- melting are almost $R_2Pd_3Ge_5$ single phase, with small amounts of R_2PdGe_6 and/or Ge. Representative microphotographs are shown in Fig. 5.2.

For the synthesis in metal flux, bismuth was initially chosen, considering some literature data on chemically affine compounds [29]. This metal solvent was found to be a reactive metal flux, forming both binary and ternary compounds, such as $LaBi_2$ (La-Pd-Ge sample); $PdBi_2$ and $Nd_{25}Pd_{25}Bi_{50}$ (Nd-Pd-Ge sample). Nevertheless, the compound of interest was only found in the La-sample in form of quite big, irregularly shaped crystals (see Fig. 5.3a).

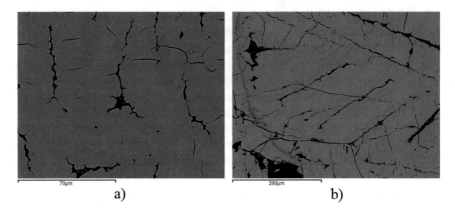

Fig. 5.2 Representative microphotographs (BSE mode) of *R*-Pd-Ge samples. **a** Ce: grey phase $Ce_2Pd_3Ge_5$; at the grain border there is a fine eutectic microstructure composed of Ce_2PdGe_6 and Ge; **b** Pr: grey phase $Pr_2Pd_3Ge_5$; at the grain border there is a fine eutectic microstructure composed of Pr_2PdGe_6 and Ge

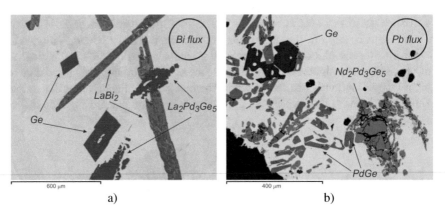

Fig. 5.3 Microphographs (BSE mode) of the La-Pd-Ge sample synthesized in Bi-flux (**a**) and of the Nd-Pd-Ge sample synthesized in Pb-flux (**b**). Reproduced with permission from [20]

In the Nd-sample the bismuth flux stabilized only the ternary compound $NdPd_2Ge_2$. Therefore, two more metal fluxes (In and Pb) were tested with Nd-Pd-Ge samples. Indium was found to mainly stabilize binary phases such as Pd_3In_7 and $NdGe_2$, with no traces of Nd-Pd-Ge ternary phases. Synthesis in Pb flux was instead successful: big crystals of $Nd_2Pd_3Ge_5$ formed, together with PdGe, re-crystallized Ge (see Fig. 5.3b) and other binary and ternary phases not visible in Fig. 5.3b. The latter was then selected for crystal separation from the flux. The sample was close in an Ar-filled quartz phial and put on a filter made by glass wool, as shown in Fig. 5.4a.

After heating the phial for one minute at 500 °C (well above the $T_m(Pb) = 327$ °C) within a resistance furnace, it was quickly removed and put into a centrifuge at a speed of 600 rpm for about 1 min. This procedure was repeated five times in order

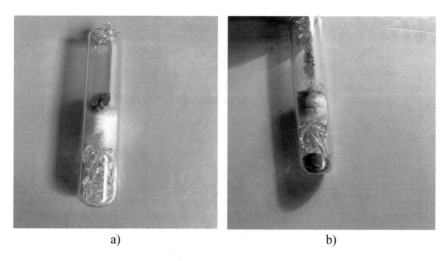

a)   b)

Fig. 5.4 Aspect of the quartz phial, containing the Nd-Pd-Ge sample synthesized with Pb-flux, before (**a**) and after (**b**) the centrifugation at 500 °C

to ensure a quantitative flux separation. At the end, remaining crystals are on the top of the glass wool whereas the lead flux lays at the bottom of the phial (Fig. 5.4b). In order to check obtained phases and the success of the separation procedure, some crystals were powdered and analysed by X-ray powder diffraction. The $Nd_2Pd_3Ge_5$, PdGe, Ge and $NdPd_2Ge_2$ phases were detected, in good agreement with SEM/EDXS analysis. In addition, the Bragg peaks for the cubic Pb were clearly revealed (Fig. 5.5), indicating an uncompleted metal flux separation.

In order to remove the remaining flux, the search for an appropriate etching solution, able to selectively oxidize residual Pb, is still under investigation.

The EDXS measured compositions of the $R_2Pd_3Ge_5$ compounds detected in the studied samples do not noticeably deviate from the ideal stoichiometry (see Table 5.2).

5.1.3 Crystal Structure of the Studied $R_2Pd_3Ge_5$ Compounds

Single crystals of $Nd_2Pd_3Ge_5$ suitable for X-ray diffraction analysis were selected from the mechanically broken sample, of nominal composition $Nd_{22.2}Pd_{11.1}Ge_{66.7}$, and glued on glass fibres for the subsequent X-ray single crystal diffraction experiment. The cell indexation was straightforward giving an orthorhombic I-centered cell. The systematic absences analysis through the recorded data set of the $Nd_2Pd_3Ge_5$ single crystal were compatible with the orthorhombic I-centered space groups $Iba2$ (№ 45) or $Ibam$ (№ 72). The preliminary structural model, easily found by *JANA2006* [30], was assumed to contain 1 Nd, 2 Pd, and 3 Ge fully occupied sites, taking into account interatomic distances and isotropic displacement parameters. The obtained

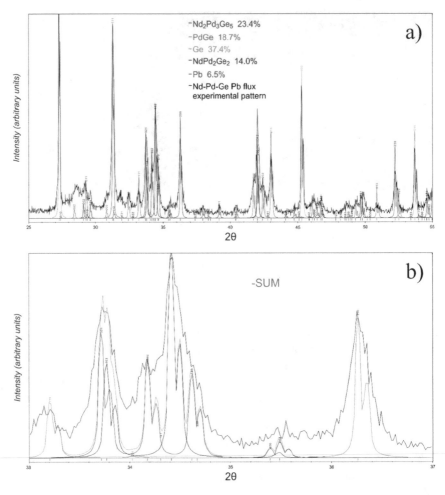

Fig. 5.5 X-ray powder pattern (**a**) for the sample Nd-Pd-Ge synthesized within a Pb flux, after centrifugation. A magnification in the range 33–37° is also shown (**b**) together with the curve obtained as a sum of the calculated diffraction patterns

$Nd_2Pd_3Ge_5$ formula is in very good agreement with the EDXS measured composition. This stoichiometric model was refined using full matrix least-squares methods with the *SHELX-14* package programs [31]. The occupancy parameters of all the crystallographic sites were varied in a separate series of least squares cycles along with the displacement parameters but they did not vary noticeably from full occupation and were assumed to be unity in further cycles. No significant residual peaks on differential Fourier maps were detected. This model was then refined anisotropically, converging to good residual values. No higher crystallographic symmetry in the tested model was found by *ADDSYM* algorithm implemented in *PLATON* [32].

Table 5.2 Compositions (measured by EDXS, at.%) and lattice parameters (from X-ray powder patterns) of the $R_2Pd_3Ge_5$ phases detected in the studied samples (* X-ray analysis was not performed)

Rare earth metal	Synthesis method	Composition			Lattice parameters [Å]			Volume [Å³]
		R	Pd	Ge	a	b	c	
La	Arc-melting	20.0	30.1	49.9	10.184(2)	12.220(3)	6.190(1)	770.4(2)
La	Bi flux*	19.9	29.3	50.8				
Ce	Arc-melting	20.5	30.6	48.9	10.1511(7)	12.131(1)	6.1568(3)	758.22(7)
Pr	Arc-melting	19.9	31.1	49.0	10.1348(5)	12.0912(8)	6.1363(2)	751.96(6)
Nd	Thermal treatment	19.2	29.7	51.1	10.125(3)	12.036(2)	6.127(1)	746.7(2)
Nd	Arc-melting	19.4	30.0	50.6	10.133(2)	12.053(3)	6.122(1)	747.7(2)
Nd	Pb flux*	19.3	31.5	49.2				
Sm	Induction melting	19.7	31.3	49.0	10.1160(9)	11.998(1)	6.0911(4)	739.28(9)

Adapted with permission from [20]

The generated CIF file has been deposited with Fachinformationszentrum Karlsruhe, 76,344 Eggenstein-Leopoldshafen, Germany (depository number CSD-431328).

Details of data collection and refinement are summarized in Table 5.3. Atomic positions along with isotropic thermal displacement parameters are listed in Table 5.4. The list of interatomic distances is reported in the Appendix (Table A5.1).

The calculated powder pattern generated from the single crystal model was used to confirm the crystal structure of the analogue 2:3:5 compounds containing La, Ce, Pr and Sm, by using X-ray diffraction results on powders. As an example, the indexed powder pattern of the almost single phase La-Pd-Ge and Nd-Pd-Ge (obtained by arc melting) samples are shown in Fig. 5.6. Other patterns are available in the Appendix (Figure A5.1). No super-reflections suggesting a possible monoclinic structure (such as 111 and 131) were observed.

The calculated cell volumes (including literature data) are plotted in Fig. 5.7 as a function of the radius of the rare earth metal, evidencing a linear decreasing, which reflects the lanthanide contraction. The trivalent radius was considered [33], taking into account literature data on similar 2:3:5 phases [26, 34].

The studied phases crystallize in the orthorhombic $oI40$-$U_2Co_3Si_5$ type, as the majority of the known $R_2T_3Ge_5$ compounds. Their crystal structure was studied in detail in the last decades and it was presented from different points of view. According to the geometrical description by Parthé [35, 36], the orthorhombic crystal structure, as well as its monoclinic derivative $mS40$-$Lu_2Co_3Si_5$, could be viewed as an intergrowth of two kinds of structural slabs, related to the $CaBe_2Ge_2$ and $BaNiSn_3$ structures respectively. More recently, some authors have interpreted the same structures as consisting of a complex three-dimensional $[T_3Ge_5]$ network spaced by the rare earth cations [37].

Table 5.3 Crystallographic data for the $Nd_2Pd_3Ge_5$ single crystal and experimental details of the structure determination. Reproduced with permission from [20]

Empirical formula	$Nd_2Pd_3Ge_5$
Structure type	$U_2Co_3Si_5$
Crystalline system	Orthorhombic
Space group	*Ibam* (№ 72)
Pearson symbol, Z	*oI*40, 4
Formula weight (M_w, g/mol)	970.63
Unit cell dimensions:	
a [Å]	10.1410(6)
b [Å]	12.0542(8)
c [Å]	6.1318(4)
V [Å^3]	749.56(8)
Calc. density (D_{calc}, g/cm^3)	8.601
Absorption coefficient (μ, mm^{-1})	40.226
Unique reflections	669 ($R_{int} = 0.0123$)
Reflections with $I > 2\sigma(I)$	651 ($R_{sigma} = 0.0276$)
Data/parameters	669/31
Goodness-of-fit on F^2 (S)	1.08
Final R indices [$I > 2\sigma(I)$]	R1 = 0.0136; wR2 = 0.0303
R indices [all data]	R1 = 0.0145; wR2 = 0.0306
$\Delta\rho_{fin}$ (max/min), [e/Å^3]	0.98/−1.03

Table 5.4 Standardized atomic coordinates and equivalent isotropic displacement parameters (Å^2) for $Nd_2Pd_3Ge_5$. Reproduced with permission from [20]

Atom	Wyckoff site	x/a	y/b	z/c	U_{iso}
Nd	8j	0.26699(2)	0.37000(2)	0	0.00721(1)
Pd1	8j	0.10788(3)	0.13975(2)	0	0.00863(8)
Pd2	4b	0.5	0	1/4	0.00904(9)
Ge1	4a	0	0	1/4	0.0089(1)
Ge2	8j	0.34381(4)	0.10632(4)	0	0.00871(9)
Ge3	8g	0	0.27498(3)	1/4	0.00856(9)

Analyzing the connectivity between the species, the latter approach can be reasonably applied also to the $R_2Pd_3Ge_5$ compounds studied here, taking the Nd compound as a representative. The Ge atoms are distributed among three independent crystallographic sites: considering only the Ge-Ge contacts at covalence range distance one can distinguish the presence of isolated Ge atoms (Ge1) and non-planar infinite –(-Ge2-Ge3-Ge2-Ge3-)$_n$– zig-zag chains. Nevertheless, the short Pd-Ge distances, ranging from 2.43 to 2.55 Å (sum of the covalent radii = 2.5 Å [38]), suggest the

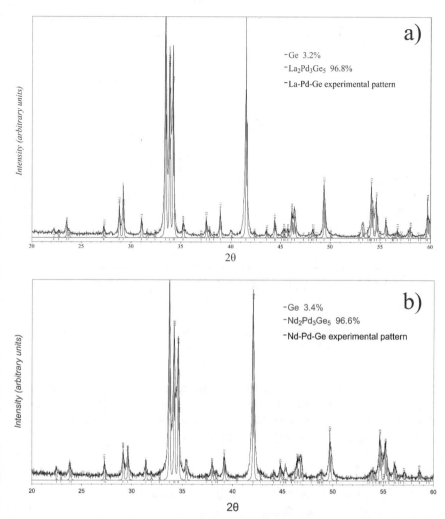

Fig. 5.6 Calculated (red/blue) and experimental (black) X-ray powder patterns for the almost single phase samples of nominal composition $R_{20.0}Pd_{30.0}Ge_{50.0}$ with $R = La$ (**a**) and $R = Nd$ (**b**)

existence of strong Pd-Ge interactions, being coherent with the idea of a $[Pd_3Ge_5]$ network (see Fig. 5.8). In this framework Pd atoms have no homocontacts (d_{min}(Pd-Pd) = 3.07 Å) and are coordinated only by Ge atoms, in two different ways: Pd1 sites are surrounded by five Ge atoms in the form of a distorted square pyramid distanced at about 2.43 (apical position), 2.49 and 2.53 Å (basal positions); Pd2 sites are surrounded by four Ge neighbours at 2.55 Å in the form of a distorted tetrahedron, with two Ge–Ge angles being 119.65° and the other two of 103.19°. Within this representation, Nd atoms are located in the biggest cavities of the three-dimensional network, being surrounded both by Ge (10 atoms) and by Pd (7 atoms)

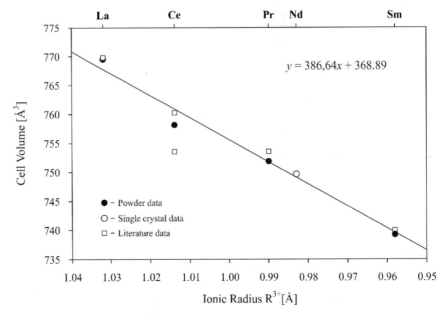

Fig. 5.7 Cell volume of $R_2Pd_3Ge_5$ compounds as a function of the R^{3+} ionic radius (literature data are taken from Refs. [11–13, 23]). Reproduced with permission from [20]

at distances ranging from 3.06 to 3.49 Å. It is interesting to note that, on the basis of the results obtained for chemical bonding in La_2PdGe_6 and Y_2PdGe_6, the presence of multiatomic Nd–Ge and polar Nd–Pd covalent bonds could be guessed and would be an interesting filed of investigation for further study.

5.1.4 Symmetry Reduction and Crystal Structure Peculiarities of R-Pd-Ge Compounds

Symmetry considerations are more and more applied for comparison of structures, since they allow us to highlight non obvious relationships, for example between structures belonging to different space groups.

These relations were definitely formalized within the group theory becoming a part of International Tables (Vol. A1) [39]. In this field, several descriptive works by Bärnighausen and Müller [40–42], accompanied by real structure examples, clearly showed how to trace out the "symmetry response" of the system with respect to the chemical/physical changes in a very concise and clear manner (the so-called "Bärnighausen tree"). According to this approach, derivative structures are described in term of symmetry reduction with respect to the most symmetrical original structure.

This approach has been used here to rationalize the crystal structure of the R-Pd-Ge ternary compounds containing 20 at.% R, among which the $LaPd_2Ge_2$, $LaPdGe_3$

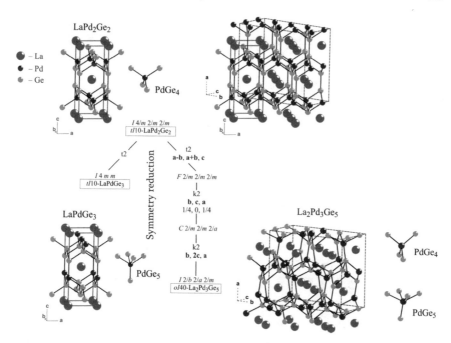

Fig. 5.8 Structural distortion associated with the symmetry reduction from $LaPd_2Ge_2$ to $LaPdGe_3$ (left branch) and to $La_2Pd_3Ge_5$ (right branch). The Pd–Ge 3D networks are indicated by black lines; the closest coordination arrangements for Pd atoms in all structures are evidenced next to each one. The metric relations between structures under consideration are also shown. Reproduced with permission from [20]

and $La_2Pd_3Ge_5$ serve as examples. Both $LaPdGe_3$ and $La_2Pd_3Ge_5$ crystal structures could be viewed as derivatives of the $LaPd_2Ge_2$ ($ThCr_2Si_2$-type) structure, which in turn is an ordered ternary derivative of the parent $BaAl_4$-type.

It is worth noting that the "$BaAl_4$ family" of related phases is one of the most populated within the intermetallics. In fact, the multitudinous investigations of the $BaAl_4$-related compounds conducted up to date revealed the existence of an incredible structural variety of them, with numerous representatives in different systems. This study is far away from being complete.

The symmetry reduction principle was applied to the $BaAl_4$-type derivatives and a beautiful structural hierarchy tree was constructed [43]. The two branches of this tree concerning the relation between $LaPd_2Ge_2$ ($ThCr_2Si_2$-type) and its 1:1:3 and 2:3:5 derivatives are reproduced in Fig. 5.8, together with the corresponding structural distortions/metrical relations of the involved structures.

In all cases the ratio between La and the other two constituent metals is 1 to 4, and the distribution of the rare earth atoms is similar. Instead, the closest arrangement around Pd atoms differs (see Fig. 5.8): in $LaPd_2Ge_2$ they are at the centers of $PdGe_4$ distorted tetrahedra (d(Pd-Ge) = 2.51 Å; Ge–Ge solid angles of 119.8° and 104.6°); in $LaPdGe_3$ palladium atoms are surrounded by five Ge atoms in the form of a

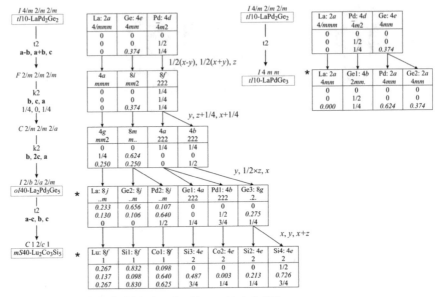

* – for the sake of clarity, the atomic positions are not standardized here, but correspond to the refined ones

Fig. 5.9 Bärnighausen tree presenting the evolution of the atomic parameters relating the LaPd$_2$Ge$_2$ with its 2:3:5 and 1:1:3 derivatives. Reproduced with permission from [20]

square pyramid (d1(Pd-Ge) $= 2.46$ Å, d2(Pd-Ge) $= 2.50$ Å), finally, in La$_2$Pd$_3$Ge$_5$ an intermediate situation occurs, with two non-equivalent Pd atoms having four and five surrounding Ge, as described above. In fact, the Pd content in the 2:3:5 compounds is intermediate between its concentration in 1:2:2 and 1:1:3 stoichiometries. The symmetry relationships between the considered structures can be highlighted through the Wyckoff positions evolution, which is shown in Fig. 5.9.

Analysing this scheme, it becomes clear that, in the derivatives, the rare earth metal suffers a negligible shift with respect to the aristotype and the different stoichiometries originate from different modes of ordered distributions of Pd and Ge species among the split Wyckoff sites. It is worth noting that the Pd and Ge atoms completely occupy their positions, without showing any tendency to statistical mixture.

In agreement with the symmetry principle [40, 41], the shortest way to obtain LaPdGe$_3$ from LaPd$_2$Ge$_2$ is a one-step *translationengleiche* (t2) decentering, resulting in the splitting of the five coordinated 4*e* site into two 2*a* independent sites. Instead, the orthorhombic La$_2$Pd$_3$Ge$_5$ is obtained from the aristotype reducing its symmetry in three steps: a *translationengleiche* (t2) decentering followed by two second order *klassengleiche* (k2) transformations, leading to the splitting of both four (4*d*) and five (4*e*) coordinated sites. A further *translationengleiche* (t2) step of symmetry reduction describes the monoclinic Lu$_2$Co$_3$Si$_5$ type distorted structure, existing for numerous R_2T_3Ge$_5$ compounds.

The described symmetry reduction steps, mainly concerning the Pd and Ge sites, are responsible for the distortion of the [Pd-Ge]$^{8-}$ polyanionic network, more pronounced for the 2:3:5 structure, also evidenced in Fig. 5.8.

5.2 The $R_2Pd_3Ge_5$ Intermetallics with Heavy Rare Earth Metals (Gd–Lu): Synthesis, Crystal Structures Refinement and Physical Properties[2]

5.2.1 Synthesis and SEM/EDXS Characterization

Samples were prepared at nominal composition $R_{20.0}Pd_{30.0}Ge_{50.0}$ ($R =$ heavy rare earth metals from Gd to Lu) in order to check the existence of the corresponding $R_2Pd_3Ge_5$ compounds. In all cases stoichiometric amounts of constituents were melted together, to obtain ingots of about 0.7 g.

Samples with $R =$ Gd, Tb, Dy, Ho, Er, Tm, Lu were prepared by repeated arc melting in Ar atmosphere on a water-cooled copper hearth with a tungsten electrode, analogously to sample with light R (La, Ce, Pr, Nd). Weight losses were less than 1%.

For the sample with $R =$ Yb, pieces of the elements were enclosed in an arc-sealed Ta crucible, in order to avoid any ytterbium evaporation, in the same way as with the Sm-containing specimen. Induction melting was performed under a stream of Ar to prevent the crucible oxidation at high temperature and it was repeated several times to ensure homogeneity. With the aim to perform different physical properties measurements, a bigger amount (about 3.6 g) of $Yb_{20.0}Pd_{30.0}Ge_{50.0}$ was obtained collecting together several samples prepared with the same procedure.

Subsequently, powders of this last sample were obtained by ball milling, performed in a Fritsch Pulverisette 7 using a stainless steel jar and 9 stainless steel balls; rotation speed was 150 rpm maintained for 1 h. The fine powders were subsequently compacted at 500 °C for 20 min under a load of 50 MPa with a vacuum condition of less than 5×10^{-2} Pa by means of a Spark Plasma Sintering (SPS) machine, PLASMAN CSP-KIT-02121 (S.S. Alloy corporation, Japan). Density of the resulting sample, determined via the conventional Archimedes method using ethanol as working liquid, was more than 94% of the theoretical value. All samples were characterized in the as-cast state.

Thermal stability of the sample with $R =$ Yb was studied by means of a differential scanning calorimeter, TG8121 (Rigaku, Japan). The measurement was carried out in the temperature range from 250 to 650 K, with a heating speed of 10 °C/min under a 200 ml/min Ar flow.

For the Yb-containing sample, a synthesis with metal flux In was performed, following thermal cycle III described in Chap. 4: 25 °C → (10 °C/min) → 750 °C

[2]Part of this section was originally published in [7].

(24 h) → (−0.5 °C/min) → 25 °C. Stoichiometric amounts of Yb, Pd and Ge, giving the nominal composition $Yb_{21}Pd_7Ge_{72}$, were put in an arc-sealed Ta crucible with a 1:45 molar excess of indium, to obtain a total mass of about 3 g.

5.2.2   Physical Properties Measurements of $Yb_2Pd_3Ge_5$

The compact pellet obtained with SPS for the Yb-containing sample was cut with a low speed diamond wheel saw, Micro Cutter MC-201 N (MARUTO corporation, Japan), obtaining slabs of rectangular (1 mm × 2 mm × 9 mm) and column (ϕ10 mm × t1.8 mm) shapes. The electrical resistivity was measured in the range 298–623 K under vacuum condition by means of a standard four-probe method. In order to check whether the studied material becomes superconducting, its resistivity was measured by means of Physical Property Measurement System (PPMS) in the range 2.2–300 K. To prevent the effects of thermal electromotive force and ionic conduction on the electrical resistivity, the AC current (0.05 Hz square wave) was used for the measurements. Seebeck coefficients were measured in the range 303–619 K under vacuum condition using a home-made apparatus. Thermal conductivity was determined from thermal diffusivity and specific heat, both measured in the 299–622 K range under a 200 ml/min Ar flow, using Laser flash analysis by LFA457 MicroFlash (NETZSCH, Germany). Magnetic susceptibility was measured at temperatures ranging from 2 to 300 K in external fields of 0.5, 1, 3, 6 T by means of a vibrating sample magnetometer on PPMS, VSM-PPMS (Quantum Design, USA). These measurements have been performed in collaboration with the Toyota Technological Institute of Nagoya and the Research Center for Environmentally Friendly Materials Engineering of Muroran, Japan.

5.2.3   On the Existence of $R_2Pd_3Ge_5$ Germanides (R = Heavy Rare Earth Metal)

The measured global compositions of analysed samples are in excellent agreement with the nominal ones. Phases identified in each sample by SEM and XRPD are listed in Table 5.5, together with the EDXS composition of the highest yield phase. The observed microstructures are regularly changing by increasing the atomic number of the R component.

For R = Gd–Ho the main phase was identified as $RPd_{2-x}Ge_{2+x}$ with tetragonal $ThCr_2Si_2$-type structure: the choice of such formula is due to the fact that the deviation of Pd and Ge content from the ideal one is remarkable. For samples with Gd and Tb, in the BSE mode SEM images this phase was observed in form of primary bright crystals, together with small amounts of R_2PdGe_6 (grey crystals) and of Ge (see Fig. 5.10a). For samples with Dy and Ho, crystals of $RPd_{2-x}Ge_{2+x}$ enclose a bright

Table 5.5 Results of SEM/EDXS and XRPD characterization of the $R_{20.0}Pd_{30.0}Ge_{50.0}$ samples. Only the sample $Yb_{20.0}Pd_{30.0}Ge_{50.0}$ (evidenced in bold) contains the 2:3:5 phase

R	Global composition (EDXS) [at.%]	Phases	Composition by (EDXS) [at.%]			Pearson symbol prototype
			R	Pd	Ge	
Gd	$Gd_{19.7}Pd_{30.0}Ge_{50.3}$	$GdPd_{2-x}Ge_{2+x}$	20.1	32.7	47.2	$tI10$-ThCr$_2$Si$_2$
		Gd_2PdGe_6	21.8	11.5	66.7	$oS72$-Ce$_2$(Ga$_{0.1}$Ge$_{0.9}$)$_7$
		Ge	-	-	100.0	$cF8$-C
Tb	$Tb_{20.5}Pd_{30.5}Ge_{49.0}$	$TbPd_{2-x}Ge_{2+x}$	20.9	35.9	43.2	$tI10$-ThCr$_2$Si$_2$
		Tb_2PdGe_6	22.8	11.6	65.6	$oS72$-Ce$_2$(Ga$_{0.1}$Ge$_{0.9}$)$_7$
		Ge	-	-	100.0	$cF8$-C
Dy	$Dy_{19.9}Pd_{29.9}Ge_{49.8}$	$DyPd_{2-x}Ge_{2+x}$	20.6	34.2	45.2	$tI10$-ThCr$_2$Si$_2$
		Dy_2PdGe_6	22.4	11.2	66.4	$oS72$-Ce$_2$(Ga$_{0.1}$Ge$_{0.9}$)$_7$
		$Dy_3Pd_{4-x}Ge_{4+x}$	26.9	30.0	43.1	$oI22$-Gd$_3$Cu$_4$Ge$_4$
		Ge	-	-	100.0	$cF8$-C
		New phase I	21.7	26.8	51.5	-
Ho	$Ho_{19.2}Pd_{29.4}Ge_{51.4}$	$HoPd_{2-x}Ge_{2+x}$	20.5	36.0	43.5	$tI10$-ThCr$_2$Si$_2$
		Ho_2PdGe_6	22.6	10.1	67.3	$oS72$-Ce$_2$(Ga$_{0.1}$Ge$_{0.9}$)$_7$
		$Ho_3Pd_{4-x}Ge_{4+x}$	27.4	30.10	42.5	$oI22$-Gd$_3$Cu$_4$Ge$_4$
		Ge	-	-	100.0	$cF8$-C
		New phase I	22.1	27.2	50.7	-

(continued)

Table 5.5 (continued)

R	Global composition (EDXS) [at.%]	Phases	Composition by (EDXS) [at.%]			Pearson symbol prototype
			R	Pd	Ge	
Er	Er$_{19.4}$Pd$_{29.9}$Ge$_{50.7}$	Er$_3$Pd$_{4-x}$Ge$_{4+x}$	28.1	31.3	40.6	oI22-Gd$_3$Cu$_4$Ge$_4$
		ErPd$_2$Ge$_2$#	19.2	22.1	58.7	
		PdGe	-	50.5	49.5	oP8-FeAs
		Ge	-	-	100.0	cF8-C
Tm	Tm$_{19.4}$Pd$_{30.3}$Ge$_{50.3}$	Tm$_3$Pd$_{4-x}$Ge$_{4+x}$	27.6	31.5	40.9	oI22-Gd$_3$Cu$_4$Ge$_4$
		TmPd$_2$Ge$_2$#	20.3	22.8	56.9	
		PdGe	-	50.1	49.9	oP8-FeAs
		Ge	-	-	100.0	cF8-C
Yb	Yb$_{20.0}$Pd$_{30.8}$Ge$_{49.2}$	**Yb$_2$Pd$_3$Ge$_5$**	**21.0**	**30.0**	**49.0**	**oI40-U$_2$Co$_3$Si$_5$**
		PdGe	-	50.2	49.8	oP8-FeAs
		Ge	-	-	100.0	cF8-C
Lu	Lu$_{18.8}$Pd$_{28.6}$Ge$_{52.6}$	Lu$_3$Pd$_{4-x}$Ge$_{4+x}$	28.0	31.3	40.7	oI22-Gd$_3$Cu$_4$Ge$_4$
		LuPdGe	34.7	31.0	34.3	oI36-YbAuSn
		PdGe	-	51.0	49.0	oP8-FeAs
		Ge	-	-	100.0	cF8-C

Phases identified only by EDXS analysis

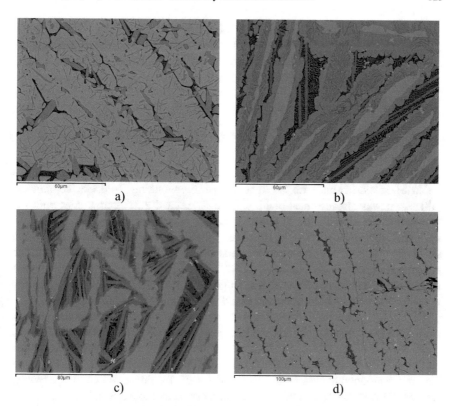

Fig. 5.10 Micrographs (SEM-BSE mode) of representative samples of $R_{20.0}Pd_{30.0}Ge_{50.0}$ nominal composition: **a** R = Gd (bright phase—$GdPd_{2-x}Ge_{2+x}$; grey phase – Gd_2PdGe_6; dark phase—Ge); **b** R = Ho (bright phase—$Ho_3Pd_{4-x}Ge_{4+x}$; grey phase—$HoPd_{2-x}Ge_{2+x}$; dark grey phase—Ho_2PdGe_6; dark phase—Ge); **c** R = Er (bright phase—$Er_3Pd_{4-x}Ge_{4+x}$; grey phase—PdGe; dark phase – Ge); **d** R = Yb (bright phase—$Yb_2Pd_3Ge_5$; grey phase—PdGe; dark phase—Ge). From [7]

core of $R_3Pd_{4-x}Ge_{4+x}$; R_2PdGe_6 crystals of elongated shape and a eutectic structure containing Ge are also observed (see Fig. 5.10b).

For the heavier rare earths Er and Tm the $RPd_{2-x}Ge_{2+x}$ does not form and $R_3Pd_{4-x}Ge_{4+x}$ is the main phase. It forms as primary bright crystals surrounded by the PdGe binary compound and a eutectic structure containing Ge (see Fig. 5.10c). In the case of Lu, the $R_3Pd_{4-x}Ge_{4+x}$ is still the highest yield phase, surrounding primary crystals of LuPdGe.

Within this series of samples, the Yb-containing alloy is unique, being almost $Yb_2Pd_3Ge_5$ single phase ($oI40$-$U_2Co_3Si_5$) with small amounts of PdGe and Ge, in agreement with XRPD results (see Fig. 5.10d), behaving very similarly to light rare earth containing alloys.

In the flux synthesized sample no traces of the wanted $Yb_2Pd_3Ge_5$ intermetallic were found. At its place, crystals of Yb_2PdGe_6, Pd_3In_7 and recrystallized Ge were detected, together with a new ternary phase of $Yb_{34.2}Pd_{17.6}Ge_{48.2}$ composition (SEM/EDXS measure). At this point, the main focus for this sample became the In flux separation and the structure characterization of this new compound. Results of this investigation are reported in Chap. 6.

5.2.4   Crystal Structure Refinement of the New $Yb_2Pd_3Ge_5$

The existence of $Yb_2Pd_3Ge_5$ was confirmed; this phase was interpreted as $oI40$-$U_2Co_3Si_5$, isostructural with the light rare earth-containing $R_2Pd_3Ge_5$ compounds [49]. As it was already mentioned, chemically related 2:3:5 germanides with the monoclinic $Lu_2Co_3Si_5$-like structure exist [4]. A one-step *translationengleiche* symmetry reduction (t2, **a-b**, **b**, **c**) relates the orthorhombic and monoclinic 2:3:5 structural models (Fig. 5.8) whose calculated diffraction patterns only differ in the presence/absence of small intensity peaks. However, a careful analysis of the good quality powder pattern of the almost single phase $Yb_2Pd_3Ge_5$ did not reveal any weak intensity super reflections associated with the monoclinic distortion.

Rietveld refinements were conducted using FullProf [44] on the diffraction powder patterns of the Yb-containing sample, prior and after the SPS treatment. In both cases, the Rietveld refinement converged with low residuals and excellent difference powder patterns. The amounts of secondary phases both in the as cast and after SPS sample are very reproducible being close to 4 and 1 wt% for PdGe and Ge, respectively. Results of the refinement after SPS are summarized in Table 5.6 and Fig. 5.11.

In the final least-squares refinement cycles 36 parameters were allowed to vary including scale factors, cell parameters, and profile parameters (Pseudo-Voigt peak shape function) for the three constituting phases. The background was refined by linear interpolation between a set of 74 background points with refinable heights. The refined positional parameters and isotropic thermal displacement parameters for all atom sites are listed in Table 5.7. Interatomic distances are listed in Table A5.2 (Appendix section).

It is interesting to note that the cell volume of $Yb_2Pd_3Ge_5$ doesn't fit well into the linear decreasing trend of the cell volumes of isostructural $R_2Pd_3Ge_5$ compounds as a function of R^{3+} ionic radius (see Fig. 5.7 and Figure A5.2). More on the Yb valence is discussed in the next paragraph.

5.2.5   Physical Properties of $Yb_2Pd_3Ge_5$

Temperature and magnetic field dependences of magnetic susceptibility χ_m and magnetization observed for $Yb_2Pd_3Ge_5$ are shown in Fig. 5.12. The temperature dependence of magnetic susceptibility (Fig. 5.12a) indicates a Curie-Weiss behavior and the magnetization linearly increases with increasing the magnetic field (Fig. 5.12b), indicating a typical paramagnetic behavior. At low temperatures, the

Table 5.6 Selected crystallographic data and structure refinement parameters for $Yb_2Pd_3Ge_5$, from [7]

Empirical formula	$Yb_2Pd_3Ge_5$
EDXS composition	$Yb_{21.0}Pd_{30.0}Ge_{49.0}$
Formula weight, M_r (g·mol^{-1})	1028.55
Space group (№)	*Ibam* (72)
Structure type	$U_2Co_3Si_5$
Pearson symbol, Z	*oI*40, 4
Angular range (°)	$5 < 2\theta < 120$
Step size (°); profile points	0.020; 5616
Unit cell dimensions:	
a, Å	10.2628(1)
b, Å	12.0580(1)
c, Å	5.98251(6)
V, Å^3	740.33(1)
Calc. density (D$_{calc}$, g·cm^{-3})	9.23
Data/Parameters	638/36
Weight fraction (%)	94.8(5)
R_B; R_F	2.15; 1.70
R_p; R_{wp}	10.00; 9.84
χ^2	2.27

magnetic susceptibility χ_m drastically increases, probably due to the presence of small amounts of Yb^{3+} containing impurities, like the ubiquitous Yb_2O_3 oxide [2, 45, 46].

For the quantitative analyses, the function fittings of the measured susceptibility were made according to the conventional relation $\chi_m = C/(T + \theta p) + \chi_0$, where χ_0 indicates the sum of Pauli paramagnetic term and diamagnetic term of ions. The resulting parameters are summarized in Table 5.8. The small, negative values of θ_P stay as another evidence of typical paramagnetism. All μ_{eff} values of $Yb_2Pd_3Ge_5$ are close to 0.8 μ_B/Yb-atom. Similar μ_{eff} were reported in the literature for other Yb-containing intermetallics, for which a nearly divalent Yb state was proposed [47, 48].

The fact that Yb valency strongly deviates from Yb^{3+} agrees well with the unit cell volume trend for $R_2Pd_3Ge_5$ compounds (see Fig. 5.7 and A5.2).

A DSC measurement preceded the electrical resistivity/thermal conductivity studies showing no peaks and thus confirming thermal stability of the compound in the 250-650 K temperature range. Figure 5.13(a) shows the temperature dependences of electrical resistivity (ρ) and Seebeck coefficient (S) for the polycrystalline $Yb_2Pd_3Ge_5$ sample.

The $\rho(T)$ shows the typical metallic-like behavior of the material with a linear increase in the whole temperature range of the measurement. Based on the PPMS measurements, the studied material is not superconducting down to 2.2 K. The

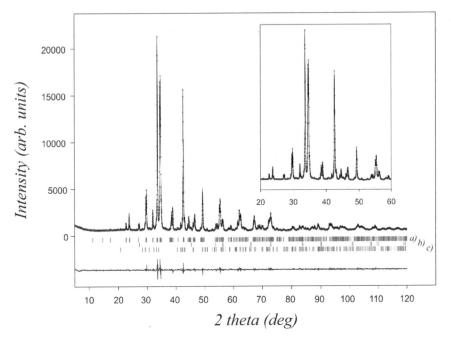

Fig. 5.11 Experimental (red dots) and calculated (continuous black line) X–ray powder diffraction patterns for the $Yb_{20.0}Pd_{30.0}Ge_{50.0}$ sample at room temperature; the lower blue line is the difference curve; vertical bars indicate the Bragg positions for $Yb_2Pd_3Ge_5$, $oI40$ (a), Ge, $cF8$ (b) and PdGe, $oP8$ (c). The inset shows a zoom of the 20–60° range. From [7]

Table 5.7 Standardized atomic coordinates and equivalent isotropic displacement parameters for $Yb_2Pd_3Ge_5$.

Atom	Site	x/a	y/b	z/c	B_{iso} (Å²)
Yb	$8j$	0.2675(1)	0.3675(1)	0	2.49(2)
Pd1	$8j$	0.1094(1)	0.1405(2)	0	2.81(4)
Pd2	$4b$	0.5	0	1/4	2.78(6)
Ge1	$4a$	0	0	1/4	2.8(1)
Ge2	$8j$	0.3408(2)	0.1023(2)	0	2.62(6)
Ge3	$8g$	0	0.2784(2)	1/4	2.25(7)

From [7]

large, positive residual resistivity ratio ($\rho_{300K}/\rho_{2.2K} = 18.7$) and the general trend of the resistivity as a function of the temperature, confirm the metallic behavior of $Yb_2Pd_3Ge_5$ and its well-ordered crystal structure. In fact, the $\rho(T)$ is described by the Bloch-Grüneisen theory [49]: it is constant below 10 K and increases with temperature, being roughly proportional to T^5 up to 50 K and to T at higher temperatures.

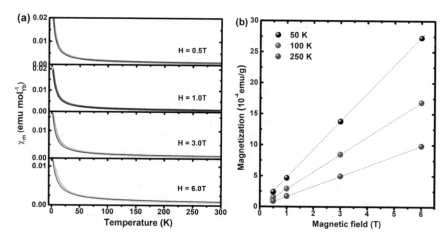

Fig. 5.12 Temperature and magnetic field dependences of (**a**) magnetic susceptibility χ_m and (**b**) magnetization for $Yb_2Pd_3Ge_5$. The black solid lines in (**a**) represent fit to the formula $\chi_m = C/(T + \theta_p) + \chi_0$. The black dash lines in (**b**) represent a linear relation of typical paramagnetic behavior. From [7]

Table 5.8 Curie-Weiss temperature (θ_P), Curie constant (C), temperature-independent orbital contribution (χ_0) and effective moment (μ_{eff}) for $Yb_2Pd_3Ge_5$. From [7]

Magnetic field (T)	θ_P (K)	C (10^{-2}emu K/mol)	χ_0 (10^{-4}emu/mol)	μ_{eff} (μ_B/Yb-atom)
0.5	−2.3	8.7	8.1	0.84
1.0	−2.3	8.4	8.0	0.82
3.0	−3.4	8.3	7.9	0.81
6.0	−6.5	9.4	6.9	0.87

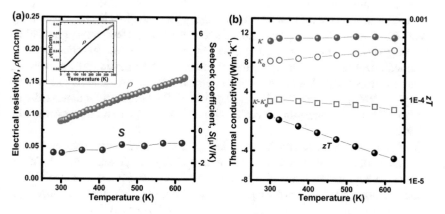

Fig. 5.13 Temperature dependences of (**a**) electrical resistivity ρ and Seebeck coefficient S, (**b**) thermal conductivity κ, electrical thermal conductivity κ_e, lattice thermal conductivity κ-κ_e, and zT dimensionless figure of merit for the polycrystalline $Yb_2Pd_3Ge_5$ sample. The inset image in (**a**) shows the ρ in the range of 2.2–300 K. From [7]

Some Yb compounds, such as Yb_2TGe_6 [50] and $Yb_{1-x}Sc_xAl_2$ [51], were reported to possess unusually large magnitude of S accompanied by low ρ. These characteristics are presumably brought about by the Yb $4f$ states near the Fermi level.

However, for $Yb_2Pd_3Ge_5$ the S value is extremely small ($-1.07 \pm 0.24\ \mu V/K$) over the whole temperature range and its negative sign implies that electrons are the majority charge carriers, confirming its metallic behavior.

Figure 5.13(b) shows the temperature dependences of thermal conductivity (κ), electrical thermal conductivity (κ_e), lattice thermal conductivity (κ-κ_e) and dimensionless figure of merit (zT) for $Yb_2Pd_3Ge_5$. The κ-κ_e, that roughly represents the lattice thermal conductivity, was calculated by using the Wiedemann Franz law: $\kappa_e = LT^{-1}$ where L is the Lorentz number (equal to $2.44 \times 10^{-8}\ W\Omega K^{-2}$ for heavily doped semiconductors or metals). In the whole range of measurement, κ remains nearly constant and the contribution of charge carrier (κ_e) is dominant due to the low magnitude of ρ, while the low contribution (κ-κ_e) of lattice might be attributed to the complex crystal structure. The obtained ρ, S, and κ define a low thermoelectric dimensionless figure of merit $zT = S^2\rho^{-1}\kappa^{-1}$ for the studied compound.

5.3 Conclusions on the $R_2Pd_3Ge_5$ Series of Intermetallics[3]

The five $R_2Pd_3Ge_5$ compounds ($R =$ La–Nd, Sm) were synthesized and characterized. Crystal structure determination/confirmation was performed on samples prepared by induction, resistance and arc-furnace melting. The use of molten metals (such as In, Bi, Pb) as solvents for the synthesis of the 2:3:5 compounds was also explored. All the tested fluxes turned out to be reactive, becoming constituents of both binary and ternary compounds. Nevertheless, large crystals of the targeted $La_2Pd_3Ge_5$ and $Nd_2Pd_3Ge_5$ stoichiometries were detected using Bi and Pb solvents respectively. These results are encouraging, since, to our knowledge, only a few ternary germanides were obtained in a similar way up to date. Considering the fact that flux may lead to the stabilization of polymorphs, the structural characterization of the obtained products is necessary. That is why, a first attempt to perform a metal flux separation through centrifugation was tested in the case of Nd–Pd–Ge sample synthesized with Pb as metal solvent along with the exploration of new $R_2Pd_3Ge_5$ representatives with heavy lanthanides.

Analyzing the connectivity between the species and literature data on related compounds, the crystal structure of the studied $R_2Pd_3Ge_5$ phases ($oI40$-$U_2Co_3Si_5$ type) was interpreted in terms of a polyanionic Pd-Ge network where also Ge–Ge covalent bonds take place. On the basis of the position-space chemical bonding analysis performed for R_2PdGe_6, the presence of R–Ge and R–Pd covalent interaction for the title compounds could be guessed, especially considering that the corresponding

---

[3]Part of this section was originally published in [7, 20].

interatomic distances in both compounds are quite similar. Nonetheless, chemical bonding studies should be carried out in order to confirm these hypotheses.

Structural relationships among R-Pd-Ge compounds (containing 20 at.% R) were searched with the help of the group-subgroup theory in the Bärnighausen formalism. The two branches of the symmetry reduction scheme concerning the relation between the 1:2:2 aristotype and its 1:1:3 and 2:3:5 hettotypes were particularly discussed, highlighting the symmetry principle effectiveness in solid state matter.

Taking into account the outlined peculiarities of Pd and Ge in the studied stoichiometric compounds, the proposed scheme can be used targeting the prediction/modelling of new possible tetrelides, corresponding to different ratios between four- and five- coordinated transition/tetrel elements.

Subsequently, the existence of the $R_2Pd_3Ge_5$ with heavy rare earth metals (R = Gd–Lu) was investigated. Alloys with nominal composition $R_{20.0}Pd_{30.0}Ge_{50.0}$ (R = Gd–Lu) were prepared by direct synthesis and characterized with respect to microstructure and phase analysis. Samples are generally multi-phase, showing different ternary compounds together with PdGe and Ge.

Instead, the Yb-containing sample is almost pure $Yb_2Pd_3Ge_5$, whose crystal structure was refined by Rietveld method obtaining low residuals. Inside the numerous $R_2T_3Ge_5$ family (R = rare earth metal; T = transition element), the studied compound is the first Yb representative crystallizing in the $oI40$-$U_2Co_3Si_5$ structure, as the already known $R_2Pd_3Ge_5$ with R = light rare earths.

In order to measure its physical properties, an $Yb_2Pd_3Ge_5$ pellet was obtained by compacting ball milled powders using SPS.

Electrical resistivity and Seebeck coefficient as a function of temperature indicate the $Yb_2Pd_3Ge_5$ metallic-like behavior, confirmed by the small zT figure of merit.

Magnetization and susceptibility measurements as a function of magnetic field and temperature indicate that the studied compound has a paramagnetic behavior and suggest a nearly divalent Yb state. The Yb^{2+} dimensions are close to those of light R^{3+} cations suggesting that the size factor plays an important role for this series of compounds.

Acknowledgements The author thanks Springer Nature and Elsevier for allowing the reuse of its own published works in this thesis.

References

1. Villars P, Cenzual K (2018/19) Pearson's crystal data. ASM International, Ohio, USA
2. Kaczorowski D, Gribanov AV, Rogl P, Dunaev SF (2016) J. Alloys Compd. 685:957–961
3. Bugaris DE, Malliakas CD, Bud'Ko SL, Calta NP, Chung DY, Kanatzidis MG (2017) Inorg Chem 56:14584–14595
4. Venturini G, Méot Meyer M, Marêché JF, Malaman B, Roques B (1986) Mat Res Bull 21:33–39
5. Ohtsu F, Fukuoka H, Yamanaka S (2009) J Alloys Compd 487:712–715
6. Freccero R, Solokha P, Proserpio DM, Saccone A, De Negri S (2017) Dalton Trans 46:14021–14033

7. Freccero R, Choi SH, Solokha P, De Negri S, Takeuchi T, Hirai S, Mele P, Saccone A (2019) J Alloys Compd 783:601–607
8. Morozkin AV, Seropegin YD, Gribanov AV, Barakatova ZM (1997) J Alloys Compd 256:175–191
9. Niepmann D, Prots YM, Pöttgen R, Jeitschko W (2000) J Solid State Chem 154:329–337
10. Hovestreydt E, Engel N, Klepp KO, Chabot B, Parthé E (1982) J Less-Common Met 85:247–274
11. Anand VK, Hossain Z, Geibel C (2008) Phys Rev B 77:184407
12. Feyerherm R, Becker B, Collins MF, Mydosh J, Nieuwenhuys GJ, Ramakrishnan R (1998) Physica B 241–243, 643–645
13. Anand VK, Thamizhavel A, Ramakrishnan S, Hossain Z (2012) J Phys Condens Matter 24:456003
14. Kurenbaeva JM, Seropegin YuD, Bodak OI, Nikiforov VN (1998) J Alloys Compd 269:151–156
15. Kurenbaeva ZM, Seropegin YD, Gribanov AV, Bodak OI, Nikiforov VN (1999) J Alloys Compd 285:137–142
16. Troc R, Wawryk R, Gribanov AV (2013) J Alloys Compd 581:659–664
17. Kaczorowski D, Belan B.D. Gladyshevskii R.R (2012) 152:839–841
18. Schmitt D, Ouladdiaf B, Routsi CD, Yakinthos JK, Seale HG (1999) J Alloys Compd 292:21–26
19. Heying B, Rodewald UC, Pöttgen R (2017) Z Kristallogr 232:435–440
20. Reproduced by permission from Springer Nature: Springer, Struct. Chem., "The $R_2Pd_3Ge_5$ (R = La–Nd, Sm) germanides: synthesis, crystal structure and symmetry reduction". Solokha P, Freccero R, De Negri S, Proserpio DM, Saccone A (2016) 27:1693–1701. ©Springer Science + Business Media New York
21. Layek S, Anand VK, Hossain Z (2009) J Magn Magn Mat 321:3447–3452
22. Singh Y, Ramakrishnan S (2004) Phys Rev B 69:174423
23. Barakatova ZM, Seropegin YD, Bodak OI, Belan BD (1995) Russ Metall 1:150–154
24. Fedyna MF, Pecharskii VK, Bodak OI (1987) Inorg Mater 23:504–508
25. Salamakha PS (1997) J Alloys Compd 255:209–220
26. Anand VK, Anupam Z, Hossain S, Ramakrishnan A, Thamizhavel DT, Adroja (2012) J Magn Magn Mat 324(16):2483–2487
27. Kaczorowski D, Pikul SP, Burkhardt U, Schmidt M, Ślebarski A, Szajek A, Werwiński M, Grin Y (2010) J Phys Condens Matter 22:215601
28. Bugaris DE, Sturza M, Han F, Im J, Chung DY, Freeman AJ, Kanatzidis MG (2015) Eur J Inorg Chem 2164–2172
29. Voßwinkel D, Pöttgen R (2013) Z. Naturforsch 68b:301–305
30. Petricek V, Dusek M, Palatinus L (2014) Z Kristallogr 229(5):345–352
31. Sheldrick GM (2008) Acta Cryst A 64:112–122
32. Spek AL (2002) PLATON. A multipurpose crystallographic tool. Utrecht University, The Netherlands
33. Shannon RD (1976) Acta Cryst 32:751
34. Chevalier B, Lejay P, Etourneau J, Vlasse M, Hagenmuller P (1982) Mat Res Bull 17:1211–1220
35. Chabot B, Parthè E, Less-Comm J (1984) Metals 97:285–290
36. Chabot B, Parthè E, Less-Comm J (1985) Metals 106:53–59
37. Zhuravleva MA, Kanatzidis MG (2003) Z. Naturforsch 58b:649–657
38. Emsley J (1999) The elements. Oxford University Press, Oxford
39. Wondratschek H, Müller U (2004) International tables for crystallography. Symmetry relations between space groups, vol A1. Academic Publishers, Dordrecht, Kluwer
40. Müller U (2006) Inorganic structural chemistry, 2 edn. Wiley
41. Müller U (2013) Symmetry relationships between crystal structures. applications of crystallographic group theory in crystal chemistry. Oxford University Press, Oxford, UK
42. Bärnighausen H (1980) Commun Math. Comput Chem 9:139–175
43. Kussmann D, Pöttgen R, UCh Rodewald, Rosenhahn C, Mosel BD, Kotzyba G, Künnen B (1999) Z Naturforsch 54b:1155–1164

44. Rodriguez-Carvajal J (2001) Recent developments in the program FullProf. Newsletter 26:12–19

45. Kowalczyk A, Falkowski M, Toliński T (2010) J Appl Phys 107:123917

46. Sales BC, Wohlleben DK (1975) Phys Rev Lett 35:1240

47. Malik SK, Adroja DT, Padalia BD, Vijayaraghavan R (1987) Magnetic susceptibility of YbCuGa, YbAgGa and YbAuGa compounds. In: Gupta LC, Malik SK (eds) Theoretical and experimental aspects of valence fluctuations and heavy fermions. Plenum Press, New York, p 463

48. Solokha P, Čurlik I, Giovannini M, Lee-Hone NR, Reiffers M, Ryan DH, Saccone A (2011) J Solid State Chem 184:2498–2505

49. Ziman JM (1960) Electrons and phonons. Oxford University Press

50. Shigetoh K, Hirata D, Avila MA, Takabatake T (2005) J Alloys Compd. 403:15–18

51. Lehr GJ, Morelli DT, Jin H, Heremans JP (2013) J Appl Phys 114:223712

Chapter 6
$Lu_5Pd_4Ge_8$, $Lu_3Pd_4Ge_4$ and Yb_2PdGe_3: Three More Germanides Among Polar Intermetallics

The R–Pd–Ge systems (R = rare earth metal) were deeply studied during the doctorate work, in particular for alloys with $at\%$Ge > 50% (see previous Chapters). In this framework, the crystal structure, chemical bonding and physical properties of some representatives were the main focuses for these investigations [1–3].

The structures of Ge-rich compounds are generally characterized by a variety of Ge covalent fragments, with topologies depending both on global stoichiometry and on the nature of the R component. These units are often joined together through Pd atoms meanwhile the R species are located in bigger channels inside the structure [4, 5], being anyway strongly involved in covalent interactions with both Ge and Pd. For these reasons, such compounds are assigned to the huge family of polar intermetallics [6] where only partial charge transfer from metals to Ge take place, allowing the formation of unusual and intriguing chemical bonding scenarios, which results in just as intriguing crystal structures.

It is interesting to remark that the ternary R–Pd–Ge compounds manifest a tendency to be stoichiometric with ordered distribution of constituents through distinct Wyckoff sites. Moreover, within the Pd–Ge fragments both species have small coordination numbers (usually 4 or 5) with very similar topological distribution of neighbours (tetrahedral coordination or its derivatives). These features may be considered as geometrical traces of a similar chemical role of Pd and Ge. That is why symmetry reduction from certain aristotypes can conveniently depict the distortions related with an ordered distribution of atom sorts. Such analysis was conducted in the literature for AlB_2 derivative polymorphs of RPdGe [7] and $BaAl_4$ derivatives of the $R_2Pd_3Ge_5$ [2, 8] family of compounds. In the systems where such type of relationships exist, the geometric factor is surely of great importance. Thus, varying R, different polymorphs [7] or even novel compounds may form. As an example, heavy rare earth containing $R_5Pd_4Ge_8$ (R = Er, Tm) [6] and $R_3Pd_4Ge_4$ (R = Ho, Tm, Yb) [5] series of compounds may be cited.

135

R. Freccero, *Study of New Ternary Rare-Earth Intermetallic Germanides with Polar Covalent Bonding*, Springer Theses, https://doi.org/10.1007/978-3-030-58992-9_6

During the exploratory syntheses conducted in the Lu–Pd–Ge and Yb–Pd–Ge systems, the Lu representatives of the abovementioned 5:4:8 and 3:4:4 stoichiometries were detected for the first time [9]. Instead, with Yb the new AlB$_2$ derivative Yb$_2$PdGe$_3$ was synthesized both by means of metal flux technique (using Indium as solvent) and direct synthesis being the first ternary germanide crystallizing with the $hP24$-Lu$_2$CoGa$_3$ structure.

In this Chapter, results on synthesis and structural characterization/analysis of these new germanides are reported. In addition, for the Lu-containing intermetallics an extensive study of their chemical bonding including Bader charges, DOS and COHP curves and a comparison with MO diagrams for Zintl anions composed by Ge is also reported. Some preliminary ELI-D and ELI-D basins results are also presented.

6.1 Synthesis, Crystal Structure and Chemical Bonding for Lu$_5$Pd$_4$Ge$_8$ and Lu$_3$Pd$_4$Ge$_4$ germanides[1]

6.1.1 Synthesis and SEM-EDXS Characterization

The Lu-Pd-Ge alloys were synthesized following different synthetic routes, including arc melting and direct synthesis in resistance furnace. In the latter case, proper amounts of components were placed in an alumina crucible, which was closed in an evacuated quartz ampoule to prevent oxidation at high temperature, and submitted to one of the following thermal cycles in a resistance furnace:

(1) 25 °C → (10 °C/min) → 950 °C (1 h) → (−0.2 °C/min) → 600 °C (168 h) → (−0.5 °C/min) → 300 °C → furnace switched off
(2) 25 °C → (10 °C/min) → 1150 °C (1 h) → (−0.2 °C/min) → 300 °C → furnace switched off

A continuous rotation of the quartz ampoule during the thermal cycle was applied. In some cases, the thermal treatment followed arc melting. Microstructure examination as well as qualitative and quantitative analyses were performed by SEM-EDXS using cobalt standard for calibration.

6.1.2 Computational Details

The electronic band structures of Lu$_5$Pd$_4$Ge$_8$ and Lu$_3$Pd$_4$Ge$_4$ were calculated by means of the self-consistent, tight-binding, linear-muffin-tin-orbital, atomic-spheres approximation method using the Stuttgart TB-LMTO-ASA 4.7 program [10] within

---

[1]Part of this section was originally published in [9].

the local density approximation (LDA) [11] of DFT. The radii of the Wigner–Seitz spheres were assigned automatically so that the overlapping potentials would be the best possible approximations to the full potential, and no empty spheres were needed to meet the minimum overlapping criterion.

The basis sets included 6 s/(6p)/5d orbitals for Lu with Lu $4f^{14}$ treated as core, 5 s/5p/4d/(4f) for palladium and 4 s/4p/(4d)/(4f) orbitals for germanium with orbitals in parentheses being downfolded.

The Brillouin zone integrations were performed by an improved tetrahedron method using a 20 × 8 × 12 k-mesh for $Lu_5Pd_4Ge_8$ and 16 × 16 × 16 for $Lu_3Pd_4Ge_4$.

Crystal Orbital Hamilton populations (COHPs) [12] were used to analyse chemical bonding. The integrated COHP values (*i*COHPs) were calculated in order to evaluate the strengths of different interactions. Plots of DOS and COHP curves were generated using wxDragon [13], setting the Fermi energy at 0 eV as a reference point.

The Bader's Quantum Theory of Atoms In Molecules (QTAIM) [14], performed by means of DGrid-4.6 [15], on the basis of FPLO calculated electron densities [30], was applied aiming to evaluate the atomic charge populations, in the title compounds. Both electron density and ELI-D were calculated thanks to a Dresden MPI-CPfS ChemBond group implementation. Calculations were done at the local density approximation (LDA) DFT level. The Perdew–Wang exchange–correlation functional [16] was used. The same k-mesh used for the previous calculation was adopted and the default convergence criteria were kept. The total electron density was generated on a discrete grid of about 0.05 Bohr.

The software Quantum Espresso [17] was used in order to relax the hypothetical "*trans*-$Lu_5Pd_4Ge_8$", containing Ge_4 units in *trans*, instead of *cis*, configuration. PBE functional for the exchange and correlation energy was used. The involved PAW sets include the $4f$, 5 s and $5p$ semicore states for Lu and the $3d$ for Ge as valence electrons. The Brillouin zone was sampled by means of a regular grid with 6 × 2 × 4 k-points. The plane wave and density cut-offs were set to 45 Ry and 450 Ry respectively, treating the orbital occupancies at the Fermi level with a Gaussian smearing of 0.01Ry.

Qualitative MO arguments based on extended Hückel theory (EHT) have been developed with CACAO package [18, 19] and its graphic interface. Even if EHT model tends to involve the most drastic approximations in MO theory, this one electron effective Hamiltonian method tends to be used to generate qualitatively correct molecular and crystal orbitals [20]. EHT is best used to provide models for understanding both molecular and solid state chemistry, as shown with great success by Roald Hoffmann and others [21].

6.1.3 Results of SEM-EDXS Characterization

The prepared samples are listed in Table 6.1, together with indication of the followed synthetic route as well as results of SEM/EDXS characterization. Information on phase crystal structure was obtained from X-ray diffraction results.

Table. 6.1 Results of SEM/EDXS characterization of the Lu–Pd–Ge samples obtained with different synthesis methods/thermal treatments

№ Overall composition [at %] Synthesis/thermal treatment	Phases	Phase composition [at %] Lu; Pd; Ge	Crystal structure
1 $Lu_{21.4}Pd_{11.2}Ge_{67.4}$ Arc melting followed by thermal treatment (1)	Lu_2PdGe_6 $Lu_5Pd_4Ge_8$ $LuPd_{0.16}Ge_2$ Ge	21.5; 12.1; 66.4 28.6; 25.1; 46.3 31.1; 5.4; 63.5 – ; –; –;	$oS72$–$Ce_2(Ga_{0.1}Ge_{0.9})_7$ $mP34$-$Tm_5Pd_4Ge_8$ $oS16$-$CeNiSi_2$ $cF8$-C
2 $Lu_{28.9}Pd_{24.1}Ge_{47.0}$ Arc melting	$Lu_5Pd_4Ge_8$ new phase Lu_2PdGe_6 $LuPd_{0.16}Ge_2$	28.8; 24.8; 46.4 32.4; 28.5; 39.1 21.7; 11.8; 66.5 30.1; 6.9; 63.0	$mP34$-$Tm_5Pd_4Ge_8$ AlB_2 related $oS72$-$Ce_2(Ga_{0.1}Ge_{0.9})_7$ $oS16$-$CeNiSi_2$
3[a] $Lu_{30.8}Pd_{25.5}Ge_{43.7}$ Direct synthesis with thermal treatment (2)	$Lu_5Pd_4Ge_8$ new phase Ge	28.6; 24.9; 46.5 33.0; 26.8; 40.2 – ; –; –;	$mP34$-$Tm_5Pd_4Ge_8$ AlB_2 related $cF8$-C
4[a] $Lu_{33.0}Pd_{26.0}Ge_{41.0}$ Arc melting followed by thermal treatment (2)	$Lu_3Pd_4Ge_4$ $Lu_5Pd_4Ge_8$ LuPdGe PdGe Ge	25.7; 35.0; 39.5 28.4; 25.1; 46.5 31.9; 34.5; 33.6 0.0; 53.4; 47.6 0.0; 0.0; 100.0	$oI22$-$Gd_3Cu_4Ge_4$ $mP34$-$Tm_5Pd_4Ge_8$ $oI36$-AuYbSn $oP8$-FeAs $cF8$-C
5 $Lu_{17.9}Pd_{29.0}Ge_{53.1}$ Arc melting	$Lu_3Pd_4Ge_4$ LuPdGe PdGe Ge	26.1; 34.2; 39.7 32.0; 33.5; 34.5 0.0; 52.4; 47.8 0.0; 0.0; 100.0	$oI22$-$Gd_3Cu_4Ge_4$ $oI36$-AuYbSn $oP8$-FeAs $cF8$-C

[a]Samples from which single crystals were taken
The highest yield phase in each sample is the first in the list, from [9]

All samples are multiphase, as it is common for non-annealed alloys belonging to complex ternary systems; Ge is always present, in some cases in small amount. SEM images in the Back-Scattered Electron (BSE) mode are well contrasted, helping to distinguish different compounds, whose compositions are highly reproducible.

Several ternary compounds already known from the literature were detected in the samples, namely Lu_2PdGe_6, $LuPd_{0.16}Ge_2$ and LuPdGe [4]. For the latter, the $oI36$-AuYbSn structure was confirmed, in agreement with previous single crystal data [7].

A new phase of composition ~ $Lu_{33}Pd_{27}Ge_{40}$ was detected in samples 2 and 3; the corresponding X-ray powder patterns can be acceptably indexed assuming a simple AlB_2-like structure, with a $\approx$ 4.28, c $\approx$ 3.54 Å, nevertheless a deeper structural investigation would be necessary to ensure its crystal structure.

Crystal structures of the new $Lu_5Pd_4Ge_8$ and $Lu_3Pd_4Ge_4$ compounds were solved by analysing single crystals extracted from samples 3 and 4, respectively. The obtained structural models, discussed in the following, are consistent with the measured powder patterns.

6.1.4 Crystal Structures of Lu₅Pd₄Ge₈ and Lu₃Pd₄Ge₄

Single crystals of $Lu_5Pd_4Ge_8$ and $Lu_3Pd_4Ge_4$ were selected from samples 3 and 4 with the aid of a light optical microscope operated in the dark field mode, and analysed by means of the four-circle Bruker Kappa APEXII CCD area-detector single crystal diffractometer.

The $Lu_5Pd_4Ge_8$ crystal selected for X-ray analysis is one more example of non-merohedral twins among germanides: previously, similar twins were found for Tb_3Ge_5 [22], Eu_3Ge_5 [23], Pr_4Ge_7 [24], La_2PdGe_6 and Pr_2PdGe_6 [1]. Basing on preliminary indexing results, the unit cell of the measured crystal might be considered as a base centered orthorhombic one with $a = 8.55$, $b = 21.29$ and $c = 13.70$ Å. The analysis of systematic extinctions suggested the following space groups: $Cmc2_1$ (№ 36), $C2cm$ (№ 40) and $Cmcm$ (№ 63). It should be mentioned that the average value of $|E^2-1| = 1.33$, characterising the distribution of peak intensities, deviates noticeably from the ideal one (0.968) for centrosymmetric space groups. Frequently, this is an indication of a twinned dataset [25, 26]. Anyway, the charge-flipping algorithm implemented in *JANA2006* [27] was used, giving a preliminary structural model with 36 Lu atoms and 96 Ge atoms in the unit cell ($Cmcm$ space group). Usually, when the scatterers have so remarkable difference of electrons, the charge-flipping algorithm is quick and very efficient in discriminating them. Considering the interatomic distances criterion and U_{eq} values, in the successive iteration cycles Pd atoms were introduced manually by substituting those of Ge, but no improvements were observed. There was no chance to improve further this model: the isotropic thermal displacement parameters showed meaningless values; several additional strong peaks were present at difference Fourier maps located too close to the accepted atom positions; the R1 value stuck at *ca.* 10%. Looking for a correct structure solution in other space groups gave no reasonable results.

At this point, a more careful analysis of diffraction spots in reciprocal space was performed using the *RLATT* [28] software. It was noticed that a remarkable amount of peaks distributed in a regular way have a small intensity and might be considered as super reflections. Therefore, they were ignored during the further indexing procedure, and a four times smaller primitive monoclinic unit cell with $a = 5.73$, $b = 13.70$, $c = 8.34$ Å and $\beta = 107.8°$ was got. The dataset was newly integrated; semi-empirical absorption corrections were applied by *SADABS* [28] software. This time, an $mP34$ structural model, containing all the atomic species, was proposed by the charge-flipping algorithm. Even so, the refinement was not satisfying: some Wyckoff sites manifested partial occupancy, it was not possible to refine the structure anisotropically, etc. It was decided to test the *ROTAX* [29] algorithm implemented in *WinGx* [30] and check the possibility to interpret our crystal as a non-merohedral twin. In fact, the two-fold rotation along the [101] direction $\begin{pmatrix} -1/2 & 0 & 1/2 \\ 0 & -1 & 0 \\ 3/2 & 0 & 1/2 \end{pmatrix}$ was proposed as a twin law obtaining a good figure of merit.

With the purpose to check this hypothesis and refine the collected data as accurately as possible, the initially selected batch of *ca.* 1000 reflections (comprising those of weak intensity considered as super reflections) was separated into two groups with the help of *CELL_NOW* [28] program suggesting the same twin law for two monoclinic domains. Successively, the information on the reciprocal domain orientation stored in the *.p4p file was used to integrate the dataset considering the simultaneous presence of both domains. After that, the resulting intensities set was scaled, corrected for absorption and merged with the help of *TWINABS* [28] program. As a result, the output in HKLF5 format with a flag indicating the original domains was generated. Using the latter and testing one more time the charge flipping procedure, the structural model was immediately found and element species were correctly assigned. The Lu$_5$Pd$_4$Ge$_8$ is of monoclinic symmetry (space group *P*2$_1$/*m*, *mP*34-Tm$_5$Pd$_4$Ge$_8$) and contains 3 Lu, 2 Pd and 6 Ge crystallographic sites. All the atom positions are completely occupied and do not manifest any considerable amount of statistical mixture. The anisotropically refined Lu$_5$Pd$_4$Ge$_8$ shows excellent residuals and flat difference Fourier maps (see Table 6.2). The refined volume ratio of twinned domains is 0.49/0.51.

Table. 6.2 Crystallographic data for Lu$_5$Pd$_4$Ge$_8$ and Lu$_3$Pd$_4$Ge$_4$ single crystals together with some experimental details of their structure determination

Empirical formula	Lu$_5$Pd$_4$Ge$_8$	Lu$_3$Pd$_4$Ge$_4$
EDXS data	Lu$_{28.6}$Pd$_{24.9}$Ge$_{46.5}$	Lu$_{25.7}$Pd$_{35.0}$Ge$_{39.5}$
Space group (№)	*P*2$_1$/*m* (11)	*Immm* (71)
Pearson symbol-prototype, Z	*mP*34-Tm$_5$Pd$_4$Ge$_8$, 2	*oI*22-Gd$_3$Cu$_4$Ge$_4$, 2
a [Å]	5.7406(3)	4.1368(3)
b [Å]	13.7087(7)	6.9192(5)
c [Å]	8.3423(4)	13.8229(9)
β (°)	107.8(1)	–
V [Å^3]	625.20(5)	395.66(5)
Abs. coeff. (μ), mm^{-1}	63.5	60.7
Twin law	$[-\frac{1}{2}0\frac{1}{2}; 0-10; \frac{3}{2}0\frac{1}{2}]$	–
k(BASF)	0.49(1)	–
Unique reflections	2105	404
Reflections I > 2σ(I)/parameters	1877/87	398/23
GOF on F^2 (S)	1.17	1.17
R indices [I > 2σ(I)]	R1 = 0.0190; wR2 = 0.0371	R1 = 0.0238; wR2 = 0.0869
R indices [all data]	R1 = 0.0247; wR2 = 0.0384	R1 = 0.0242; wR2 = 0.0871
$\Delta\rho_{fin}$ (max/min), [e/Å^3]	2.00 / –2.83	2.87 / –3.33

From [9]

The *RLATT* program was used to generate a picture showing the distribution of X-ray diffraction spots originated from the two domains differentiated by colour in Fig. 6.1 (upper part). The distribution of the non-overlapped peaks of the second domain is also well visible on the precession photo of the *h3l* zone demonstrated in Fig. 6.1 (lower part). In the same figure, a schematic real space representation of the mutual orientation of the twinned-crystal components is shown. Selected crystallographic data and structure refinement parameters for the studied single crystals are listed in Table 6.2.

The indexing of the diffraction dataset of the $Lu_3Pd_4Ge_4$ single crystal gives an orthorhombic base centered unit cell with $a = 4.137$, $b = 6.919$, $c = 13.823$ Å. Systematic extinction conditions related to symmetry elements presence were not found for this dataset. The structure solution was found in *Immm* with the aid of the

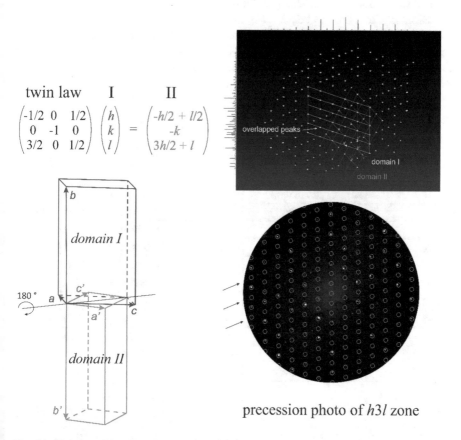

Fig. 6.1 Twin law and reciprocal orientation of the two domains in the $Lu_5Pd_4Ge_8$ twinned crystal (left); distribution of the diffraction peaks in the reciprocal space (right). Nodes of the reciprocal pattern for each domain are shown in white and green, overlapped peaks are yellow. On the experimental precession photos of the *h3l* zone, arrows indicate the directions along which the second domain peaks are well visible. From [9]

charge flipping algorithm implemented in *JANA2006* [27]. The proposed preliminary structural model contains five crystallographic sites, giving the $Lu_3Pd_4Ge_4$ formula and corresponding to the $oI22$-$Gd_3Cu_4Ge_4$ prototype. Partial site occupation (due to a possible statistical mixture of the species) was checked in separate cycles of least-squares refinement, but no significant deviation from fully occupation was detected. The final structure model was refined as stoichiometric with the anisotropic displacement parameters for all crystallographic sites, giving small residual factors and a flat difference Fourier map (see Table 6.2). The standardized atomic coordinates for $Lu_5Pd_4Ge_8$ and $Lu_3Pd_4Ge_4$ are given in Table 6.3.

At the end of the crystal structure solution procedure, the generated CIF files have been deposited at Fachinformationszentrum Karlsruhe, 76,344 Eggenstein-Leopoldshafen, Germany, with the following depository numbers: CSD-434226 ($Lu_5Pd_4Ge_8$) and CSD-434225 ($Lu_3Pd_4Ge_4$).

Similarly to $(Tm/Er)_5Pd_4Ge_8$ [6] the presence of Ge covalent fragments in $Lu_5Pd_4Ge_8$ is obvious on the basis of interatomic distances analysis. Among these, there are two almost identical Ge–Ge dumbbells distanced at 2.49 Å and one more finite fragment composed of four germanium atoms having a *cis*-configuration ("*cis*-Ge_4") (Fig. 6.2, Table 6.4).

Table. 6.3 Atomic coordinates standardized by Structure Tidy [31] and equivalent isotropic displacement parameters for $Lu_5Pd_4Ge_8$ and $Lu_3Pd_4Ge_4$

Atom	Site	x/a	y/b	z/c	U_{eq} (Å^2)
$Lu_5Pd_4Ge_8$					
Lu1	$2e$	0.71858(8)	0.25	0.93028(6)	0.0047(1)
Lu2	$4f$	0.13606(7)	0.11370(2)	0.78913(7)	0.0051(1)
Lu3	$4f$	0.62176(8)	0.11902(2)	0.28943(7)	0.0056(1)
Pd1	$4f$	0.07436(13)	0.08476(3)	0.14089(12)	0.0072(1)
Pd2	$4f$	0.42601(13)	0.58211(3)	0.35985(12)	0.0075(1)
Ge1	$2e$	0.0515(2)	1/4	0.28977(15)	0.0081(2)
Ge2	$2e$	0.3343(2)	1/4	0.58221(15)	0.0078(2)
Ge3	$2e$	0.7797(2)	1/4	0.5814(2)	0.0063(2)
Ge4	$4f$	0.15453(17)	0.04252(5)	0.44776(16)	0.0071(1)
Ge5	$2e$	0.2797(2)	1/4	0.0606(2)	0.0048(2)
Ge6	$4f$	0.34622(17)	0.54443(4)	0.05049(16)	0.0060(1)
$Lu_3Pd_4Ge_4$					
Lu1	$2a$	0	0	0	0.0110(2)
Lu2	$4j$	1/2	0	0.37347(4)	0.0081(2)
Pd	$8l$	0	0.30094(10)	0.32738(5)	0.0155(3)
Ge1	$4h$	0	0.18745(17)	0.5	0.0084(3)
Ge2	$4i$	0	0	0.21754(10)	0.0132(3)

From [9]

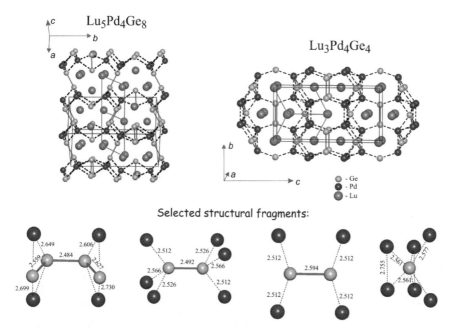

Fig. 6.2 Crystal structures of $Lu_5Pd_4Ge_8$ and $Lu_3Pd_4Ge_4$. The Pd–Ge frameworks are evidenced by dotted lines. Ge–Ge covalent bonds are shown by red sticks. Selected fragments, discussed in the text, are pictured at the bottom. Selected interatomic distances (Å) are indicated. $ThCr_2Si_2$-like fragments are evidenced in blue. From [9]

The latter manifests a small geometrical distortion from the ideal conformation due to slightly different chemical arrangements around terminal Ge atoms (terminal atoms are located at 2.56 and 2.63 Å far from central dumbbell; the internal obtuse angles are *ca.* 111° and 113°, respectively). The *cis* unit is planar and lays at the mirror plane of the $P2_1/m$ space group. The cited covalent fragments are joined together through Pd–Ge contacts shortened with respect to metallic/atomic radii sum (ranging from 2.51 to 2.73 Å) in a complex network hosting Lu atoms in the biggest cavities (see Fig. 6.2). The shortest Lu–Pd and Lu–Ge contacts do not manifest noticeable deviations from the values of *ca.* 3.0 Å, similarly to what was revealed for the same distances in R_2PdGe_6 and $R_2Pd_3Ge_5$ phases, suggesting some chemical analogy.

The $Lu_3Pd_4Ge_4$ contains a lower amount of germanium with respect to $Lu_5Pd_4Ge_8$, as a consequence, only a simple Ge–Ge dumbbell forms being, however, more stretched (2.59 Å, Table 6.5). The trend of other interactions is similar as for $Lu_5Pd_4Ge_8$: Pd and Ge construct an extended network with infinite channels of hexagonal and pentagonal form hosting Lu atoms.

One more structural relation can be proposed for title compounds: both of them contain common structural $ThCr_2Si_2$-like building blocks [32] (highlighted in blue line in Fig. 6.2) defined in many related compounds as "linkers" within various polyanionic fragments [33].

Table 6.4. Interatomic distances and integrated crystal orbital Hamilton populations (–iCOHP, eV/cell) at E$_F$ for the strongest contacts within the first coordination spheres in Lu$_5$Pd$_4$Ge$_8$

Central atom	Adjacent atoms	d (Å)	-iCOHP	Central atom	Adjacent atoms	d (Å)	-iCOHP	Central atom	Adjacent atoms	d (Å)	– iCOHP
Lu1	Ge6 (×2)	2.853	1.26	**(1b)Ge3**	Ge5	2.904	1.25	**(1b)Ge6**	Ge6	2.494	2.39
	Ge1	3.025	0.81		Ge3	2.938	1.25		Pd1	2.516	2.16
	Ge5	3.033	1.02		Pd1	3.042	0.71		Pd2	2.533	2.12
	Ge3	3.036	0.85		Ge1	3.051	0.92		Pd1	2.619	1.80
	Ge2	3.064	0.79		Pd2	3.063	0.68		Lu1	2.853	1.26
	Ge5	3.069	0.81		Ge6	3.072	0.90		Lu2	3.016	0.96
	Pd1 (×2)	3.194	0.52		Ge6	3.094	0.75		Lu2	3.050	0.80
	Pd2 (×2)	3.258	0.46		Pd2	3.100	0.63		Lu3	3.072	0.90
Lu2	Ge5	2.857	1.29		Ge4	3.105	0.71		Lu3	3.094	0.75
	Ge3	2.918	1.20		Ge4	3.120	0.84	**Pd1**	Ge6	2.516	2.16
	Ge2	2.994	0.99		Pd1	3.236	0.52		Ge4	2.526	2.11
	Ge6	3.016	0.95		Ge4	3.493	0.27		Ge1	2.606	1.66
	Ge4	3.043	0.91	**(1b)Ge4**	Ge2	2.559	1.92		Ge6	2.619	1.80
	Ge4	3.043	0.79		Pd2 (×2)	2.699	1.46		Ge5	2.730	1.35
	Ge6	3.050	0.80		Lu2 (×2)	2.918	1.20		Lu3	3.042	0.71
	Pd1	3.087	0.68		Lu3 (×2)	2.938	1.25		Lu2	3.087	0.68
	Pd1	3.104	0.63		Lu1	3.036	0.86		Lu2	3.104	0.63
	Pd2	3.114	0.64		Ge4	2.492	2.48		Lu1	3.194	0.52
	Pd2	3.156	0.57		Pd2	2.512	2.14	**Pd2**	Ge4	2.512	2.14
(2b)Ge1	Ge2	2.484	2.98		Pd1	2.526	2.11		Ge6	2.533	2.12
	Pd1 (×2)	2.606	1.66		Pd2	2.566	1.94		Ge4	2.566	1.94

(continued)

Table. 6.4. (continued)

Central atom	Adjacent atoms	d (Å)	-iCOHP	Central atom	Adjacent atoms	d (Å)	-iCOHP	Central atom	Adjacent atoms	d (Å)	– iCOHP
	Ge5	2.627	1.69		Lu2	3.043	0.91		Ge2	2.649	1.53
	Lu1	3.025	0.81		Lu2	3.043	0.79		Ge3	2.699	1.46
	Lu3 (×2)	3.051	0.92		Lu3	3.105	0.71		Lu3	3.063	0.68
(2b)Ge2	Ge1	2.484	2.98	(1b)Ge5	Ge1	2.627	1.69		Lu3	3.100	0.63
	Ge3	2.559	1.92		Pd1 (×2)	2.730	1.35		Lu2	3.114	0.65
	Pd2 (×2)	2.649	1.53		Lu2 (×2)	2.857	1.29		Lu2	3.156	0.58
	Lu2 (×2)	2.994	1.00		Lu3 (×2)	2.904	1.26				
	Lu1	3.064	0.79		Lu1	3.033	1.02				
					Lu1	3.069	0.81				

Symbols (2b) and (1b) indicate the number of homocontacts for corresponding Ge species, for more details on it see next paragraph. From [9]

Table. 6.5. Interatomic distances and integrated crystal orbital Hamilton populations (-iCOHP, eV/cell) at E$_F$ for the strongest contacts within the first coordination spheres in Lu$_3$Pd$_4$Ge$_4$

Central atom	Adjacent atoms	d (Å)	-iCOHP	Central atom	Adjacent atoms	d (Å)	-iCOHP
Lu1	Ge4 (×4)	2.992	1.21	**(0b)Ge2**	Pd (×4)	2.562	1.88
	Ge5 (×2)	3.006	1.05		Pd (×2)	2.577	1.86
	Pd (×8)	3.445	0.41		Lu2 (×2)	2.988	0.83
Lu2	Ge5 (×2)	2.988	0.83		Lu1	3.006	1.05
	Ge4 (×4)	3.003	0.99	**Pd**	Ge4	2.512	2.23
	Pd (×4)	3.003	0.79		Ge5 (×2)	2.562	1.88
	Pd (×2)	3.100	0.58		Pd	2.755	0.97
(1b)Ge1	Pd (×2)	2.512	2.23		Lu2 (×2)	3.003	0.79
	Ge1	2.595	1.82		Pd (×2)	3.058	0.46
	Lu1 (×2)	2.992	1.22		Lu2	3.100	0.58
	Lu2 (×4)	3.003	0.99		Lu1 (×2)	3.445	0.41

Symbols (1b) and (0b) indicate the number of homocontacts for corresponding Ge species, for more details on it see next paragraph. From [9]

6.1.5 Lu$_5$Pd$_4$Ge$_8$: Structural Relationships

Looking for structural relationships is not an easy task, since this process is often strongly affected by the human factor and it is based on doubtful criteria. From this point of view, one of the most rigorous approaches is based on the symmetry principle within the group-subgroup theory [34]. The most frequent chemical reason causing the reduction of symmetry is the so-called "colouring", which can be interpreted as an ordered distribution of different chemical elements within distinct Wyckoff sites. Müller [35] and Pöttgen [36] depict numerous examples of these (and it will be also the case for Yb$_2$PdGe$_3$, discussed in Sect. 6.3).

Structural relationships between Tm$_5$Pd$_4$Ge$_8$ (isostructural with Lu$_5$Pd$_4$Ge$_8$) and R_3T_2Ge$_3$ (T = late transitional element) were proposed in the literature [6] based on topological similarities between polyanionic fragments and spatial distribution of cations. An alternative description of relationships between the abovementioned structures in terms of symmetry reduction is proposed here. The stoichiometries of these compounds may be related as follows:

$$4\,R_3T_2\text{Ge}_3 - 2\,R + 4\,\text{Ge} = 2\,R_5T_4\text{Ge}_8$$

This relation, even if purely numerical, finds support when comparing the crystal structures of the two chemically affine representatives Lu$_3$Fe$_2$Ge$_3$ (oS32) and Lu$_5$Pd$_4$Ge$_8$ (mP34). As it is evidenced in Fig. 6.3, one of the Lu sites in the former is substituted by a Ge dumbbell in the latter.

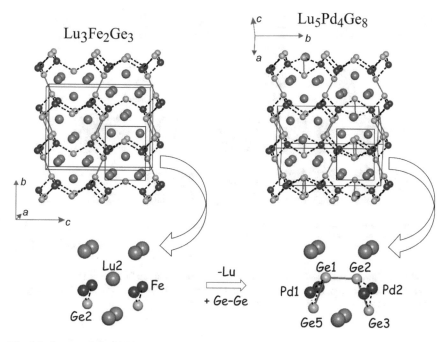

Fig. 6.3 Structural similarities between $Lu_3Fe_2Ge_3$ and $Lu_5Pd_4Ge_8$. The polyanionic networks are shown by dotted lines, covalent Ge fragments are joined by red sticks. The grey rectangle evidences regions of the crystal space where Lu/Ge2 substitution takes place (for details see text). From [9]

From the chemical interaction point of view this should be a drastic change, instead, the remaining atoms apparently do not suffer noticeable displacements. That is why it was checked whether a Bärnighausen tree might be constructed relating the $oS32$ and $mP34$ models. In fact, only two reduction steps are needed:

- a *traslationengleiche* (t2) decentering leading to a monoclinic Niggli cell ($mP16$-$P2_1/m$).
- a *klassengleiche* transformation (k2) giving a monoclinic model with doubled cell volume ($mP32$-$P2_1/m$).

As a result, all the independent sites split in two (see Fig. 6.4).

The Lu2' site ($2e$: $0.211\ 1/4\ 0.430$) is further substituted by two germanium atoms (positions Ge1 and Ge2 in the final $mP34$-$P2_1/m$ structural model). As a result, the already cited *cis*-Ge4 unit forms (see Fig. 6.2), whose chemical role is discussed in the next paragraph. The presence of the *cis*-Ge$_4$ units is quite intriguing, since the *trans* conformation is more favorable in numerous molecular chemistry examples. So, it was decided to generate a structural model of $Lu_5Pd_4Ge_8$ composition hosting the *trans*-Ge$_4$ unit. This task is quite straightforward, considering the low symmetry of the title compound and the fact that the "cis" unit lays on the mirror plane of the $P2_1/m$ space group. These structural changes are schematically represented in Fig. 6.5 and the obtained atomic parameters for the new structural model (further

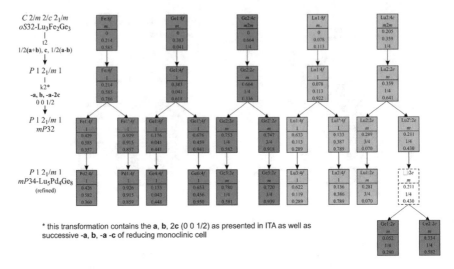

Fig. 6.4 Evolution of the atomic parameters within the Bärnighausen formalism accompanying the symmetry reduction from Lu$_3$Fe$_2$Ge$_3$ to Lu$_5$Pd$_4$Ge$_8$ structures. The background colors correspond to the atom markers in the figures through the text. From [9]

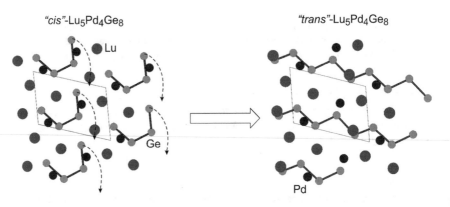

Fig. 6.5 Schematic representation of the structural relationships between "cis-" and "trans-Ge$_4$" fragments in Lu$_5$Pd$_4$Ge$_8$ models. It is important to note that, moving to the trans conformation, infinite zigzag Ge chains form

called "trans-Lu$_5$Pd$_4$Ge$_8$") are listed in Table 6.6. It is to be noted that, changing the conformation of Ge$_4$ from "cis" to "trans", zigzag Ge chains form.

Successively, the "trans-Lu$_5$Pd$_4$Ge$_8$" was relaxed with the aid of Quantum Espresso in 40 relaxation steps (ca. 25 iteration each). Surprisingly, the optimized model perfectly coincides with the experimentally found "cis-Lu$_5$Pd$_4$Ge$_8$", confirming that the minimal energy is associated with the latter. An animation showing the evolution of the structural model along with each relaxation step, e.g. from "trans" to "cis", was also generated.

Table. 6.6 Cell and atomic parameters generated for the "trans"- $Lu_5Pd_4Ge_8$ model

Space-group	$P2_1/m$ (11)—monoclinic			
Cell	$a = 5.7406$ Å $b = 13.7087$ Å $c = 8.3423$ Å $\beta = 107.77°$ $V = 625.18$ Å^3 $Z = 2$			
Atom	Wyck	x/a	y/b	z/c
Ge1	2e	0.0515	1/4	0.2897
Ge2	2e	0.3343	1/4	0.5822
Ge3	4f	0.1545	0.0425	0.4477
Ge4	2e	0.2797	1/4	0.0606
Ge5	4f	0.3462	0.5444	0.0505
Ge6	2e	0.9363	3/4	0.1910
Pd1	4f	0.0743	0.0847	0.1409
Pd2	4f	0.4259	0.5821	0.3598
Lu1	2e	0.7185	1/4	0.9302
Lu2	4f	0.1360	0.1137	0.7891
Lu3	4f	0.6217	0.1190	0.2894

6.1.6 Chemical Bonding Analysis

Frequently, chemical bonding in polar intermetallics is preliminary addressed using the Zintl-Klemm concept. Taking into account the interatomic distances between Ge atoms the presence of $[(1b)Ge^{3-}]$ with $[(2b)Ge^{2-}]$ Zintl species in $Lu_5Pd_4Ge_8$ and $[(1b)Ge^{3-}]$ with $[(0b)Ge^{4-}]$ ones in $Lu_3Pd_4Ge_4$ could be guessed. In order to guarantee the precise electron count, an average number of valence electrons per Ge atom $[VEC(Ge)]$ should amount to 6.75 for $Lu_5Pd_4Ge_8$ and to 7.50 for $Lu_3Pd_4Ge_4$. Although it is reasonable to hypothesize a formal charge transfer of 3 valence electrons per Lu atom (Lu^{3+}), to a first approximation, the Pd could be considered, formally, as a divalent cation (Pd^{2+}) or a neutral species (Pd^0). However, none of the possible electron distribution formula listed below is suitable for studied compounds, giving $VEC(Ge)$ that somewhat deviates from ideal values.

$Lu_5Pd_4Ge_8$ (Pd^0)	$VEC(Ge) = 5.875$
$Lu_5Pd_4Ge_8$ (Pd^{2+})	$VEC(Ge) = 6.875$
$Lu_3Pd_4Ge_4$ (Pd^0)	$VEC(Ge) = 6.250$
$Lu_3Pd_4Ge_4$ (Pd^{2+})	$VEC(Ge) = 8.250$

Even if the obtained $VEC(Ge)$ are closer to 6.75/7.50 in the case of Pd^{2+}, this assumption is not coherent with the valence electrons flow when considering any of the known electronegativity scales. For example, taking into account the *Pearson* electronegativity values for Pd (4.45 eV) and Ge (4.60 eV) it is clear that a charge transfer from Pd to Ge is hardly probable. Strictly speaking, it is not possible to successfully apply the (8–N) rule in order to interpret the Ge–Ge covalent interactions. Thus, it becomes clear that these simplified considerations are not sufficient to account for chemical bonding in the studied intermetallics. In particular, it is not reliable to consider covalent Ge fragments as isolated and more complex interactions should be taken into account, similarly to what was already highlighted in previous Chapters for R–Pd–Ge intermetallics. Therefore, a deeper chemical bonding investigation was conducted.

In Table 6.7 the volumes of the atomic basins and Bader effective charges for all the atoms in Lu$_5$Pd$_4$Ge$_8$ and Lu$_3$Pd$_4$Ge$_4$ are listed together with those for the same species in their pure elements form. Comparing these values, one can qualitatively estimate the chemical role of constituents in binary/ternary compounds.

In both ternary germanides, the QTAIM basins of Lu are shrunk with respect to Lu-$hP2$, and the corresponding charges oscillate around +1.4, confirming the active metal-like role of Lu. Anyway, the significant difference between Lu effective charges and the formal one (+3) suggest that some of its valence electrons may contribute to covalent interactions with both Ge and Pd.

The palladium atoms have similar volumes of atomic basins (*ca.* 20 Å^3) and are negatively charged (~ −0.7) suggesting a bonding scenario coherent with the electronegativity values supporting the idea that Pd takes part in a polyanionic network as it was hypothesized from crystal structure analysis. It is worth to note that in the same compound Ge atoms have pronounced differences of the charge values (always negative), from site to site. More on structural/chemical reasons for that is discussed in the following.

Table. 6.7 Calculated QTAIM effective charges and atomic basin volumes for Lu, Pd and Ge in their elemental structure, in Lu$_5$Pd$_4$Ge$_8$ and in Lu$_3$Pd$_4$Ge$_4$

Element/compound	Atom/site	Volume, [Å^3]	QTAIM charge, Q^{eff}	Compound	Atom/site	Volume, [Å^3]	QTAIM charge, Q^{eff}
Lu (*hP2*)	Lu/2c	29.74[a]	0	Lu$_5$Pd$_4$Ge$_8$ (*mP34*)	Lu1/2e	15.65	+ 1.45
Pd (*cF4*)	Pd/4a	14.71[a]	0		Lu2/4f	15.27	+ 1.42
Ge (*cF8*)	Ge/8a	22.66[a]	0		Lu3/4f	15.60	+ 1.43
Lu$_3$Pd$_4$Ge$_4$ (*oI22*)	Lu1/2a	16.44	+ 1.35		Pd1/4f	19.21	− 0.70
	Lu2/4j	14.78	+ 1.40		Pd2/4f	19.08	− 0.71
	Pd/8 *l*	18.96	− 0.66		(2*b*)Ge1/2e	18.97	− 0.19
	(1*b*)Ge1/4 *h*	21.86	− 0.92		(2*b*)Ge2/2e	19.03	− 0.23
	(0*b*)Ge2/4i	16.14	− 0.18		(1*b*)Ge3/2e	22.11	− 0.84
					(1*b*)Ge4/4f	18.34	− 0.33
					(1*b*)Ge5/2e	23.40	− 1.10
					(1*b*)Ge6/4f	19.23	− 0.55

[a] The QTAIM volumes of atoms in pure elements are equal to the volumes of their Wigner–Seitz polyhedra; structural data were taken from [1]

The total and projected DOS for Lu, Pd and Ge for the studied intermetallics are shown in Fig. 6.6. The orbital projected DOS can be found in the Appendix (Figure A6.1). Focusing on the total DOS, a difference between the two compounds at the Fermi energy (E_F) is evident: for $Lu_5Pd_4Ge_8$ a pseudo-gap is visible just above E_F, instead for $Lu_3Pd_4Ge_4$ the Fermi level corresponds to a local maximum of the DOS, indicating a potential electronic instability. This might be a sign of particular physical properties (e.g. superconductivity or magnetic ordering) [37] or of small structural adjustments (e.g. off-stoichiometry due to statistical mixture or increase of vacancy concentration) [38] which, adequately modelled, would shift the E_F towards a local minimum. Even if EDXS elementary composition is compatible with a slightly off-stoichiometry, there is no strong indication of this coming from XRD data, so, the stoichiometric model was considered here. Further experimental investigations are already planned in order to investigate physical properties of $Lu_3Pd_4Ge_4$.

For both compounds, a valence orbital mixing of the three components over the whole energy range is noteworthy. Below E_F, both DOSs show a gap around -7 eV separating two regions, the lowest being mostly dominated by the *4 s* Ge states.

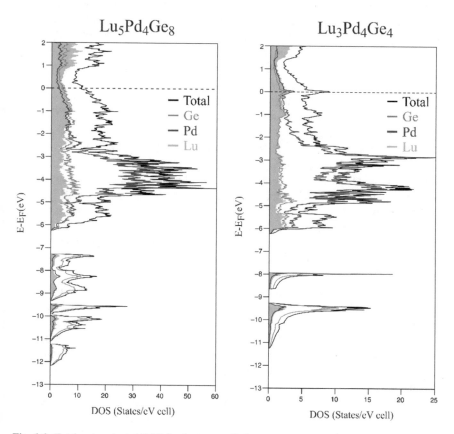

Fig. 6.6 Total and projected DOS for the two studied compounds. From [9]

The Pd-d states are mainly distributed in the range between -5 and -2.5 eV; their width and energy overlap with $4p$ Ge and Lu states support the bonding relevance of Pd–Ge and Pd–Lu interactions. The fact that the major part of Pd $4d$ states is located well below the E$_F$ indicates the electron acceptor character of this species. A significant contribution of $5d$ Lu states just below the E$_F$ is a common feature for cations in polar intermetallics characterized by an incomplete charge transfer (in good agreement also with Bader's charge values).

Although the *Zintl-Klemm* (8–N) rule cannot be applied for the title compounds, it was decided to trace interaction similarities making a comparative discussion on the electronic structures of ideal Zintl anions Ge$_2^{6-}$ and *cis*-Ge$_4^{10-}$ coming from extended Hückel calculation with those obtained by means of TB-LMTO-ASA in terms of COHP curves. Molecular orbital diagrams (MO) for Ge$_2^{6-}$ (point group $D_{\infty h}$) and Ge$_4^{10-}$ (the point symmetry of this anion was forced to C_{2v} point group fixing all the distances to 2.56 Å and obtuse internal angles to 111°) are presented in the Appendix (Figure A6.2) with the accordingly labeled orbitals.

In Fig. 6.7a the molecular orbital overlap population (MOOP) for Ge$_2^{6-}$ is shown together with COHP curves for Ge–Ge interactions (in dumbbells) existing in Lu$_3$Pd$_4$Ge$_4$ and Lu$_5$Pd$_4$Ge$_8$. It is known that these partitioning methods could not be directly compared since MOOP partitions the electron number, instead, COHP partitions the band structure energy. In addition, they were obtained at different level of theory: EHT and DFT, respectively. However, since they both permits to easily distinguish between bonding and antibonding states, it was decided to perform a qualitative comparison targeting to figure out the similarities/differences between isolated molecular fragments analogous with those found in the compounds studied.

The presence of the gap (at *ca.* -7 eV) may be attributed to the energy separation between the σss and σ^*ss of Ge fragments from the σp, πp and π^* orbitals. For the Ge dumbbells in Lu$_3$Pd$_4$Ge$_4$ there are some occupied π^* states close to E$_F$, whereas in Lu$_5$Pd$_4$Ge$_8$, the cited interactions are almost optimized at E$_F$. Anyway, in both cases, less antibonding states are populated then in the hypothesized Ge-Zintl molecular anions. Since, for Ge$_2^{6-}$ and Ge$_4^{10-}$ the considered antibonding states can be associated to lone pairs, this observation lead to the conclusion that within title intermetallics "lone pairs" are used to establish some more covalent interactions with the surrounding metal atoms. From these observations it derives that Ge dumbbells and the *cis*-Ge$_4$ fragment are not completely polarized.

From the structural data it is known that in Lu$_3$Pd$_4$Ge$_4$, Ge atoms within dumbbells are distanced at 2.59 Å as in diverse metal-like salts studied before [39–41]. Instead, in Lu$_5$Pd$_4$Ge$_8$ this distance is shortened to 2.49 Å. Usually, the trend of Ge–Ge dumbbell distances is related with electrostatic repulsion between atoms. This statement is coherent with integrated COHP values ($-i$COHP, see Tables 6.4 and 6.5) reflecting the same trend, being of -1.82 eV/cell for Lu$_3$Pd$_4$Ge$_4$ and of -2.39 and -2.48 eV/cell for Lu$_5$Pd$_4$Ge$_8$.

Within the *cis*-Ge$_4^{10-}$ anion the number of covalent interactions is higher; as a result the energy dispersion of its molecular states increases. For example, in the range -18 to -14 eV there are four MOs instead of two MOs for dumbbells. A very similar trend/type of interactions derives from COHP curves for Lu$_5$Pd$_4$Ge$_8$. Based

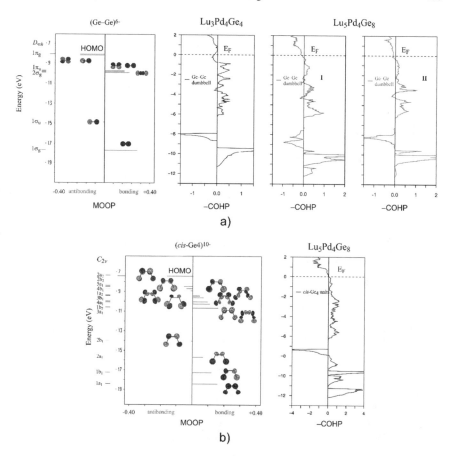

Fig. 6.7 Extended Hückel calculated Molecular Orbital Overlap Population (MOOP) plot for the Ge_2^{6-} (**a**) and cis-Ge_4^{10-} (**b**) anions together with the corresponding Crystal Orbital Hamilton Population (COHP) for $Lu_3Pd_4Ge_4$ and $Lu_5Pd_4Ge_8$ (I and II corresponds to two distinct dumbbells). The degeneracy of the π levels for Ge_2^{6-} is removed for sake of clarity. The HOMO energy is set in correspondence of the E_F. From [9]

on –iCOHP values listed in Tables 6.4 and 6.5 it derives that Pd–Ge interactions are very relevant, so one may assume the covalent type of bonding between them. The—COHP plots in Fig. 6.8 confirm that they are mainly of bonding type over a large range below E_F with a weak unfavorable antibonding interaction in the vicinity of E_F, probably due to electrostatic repulsion between Ge orbitals and filled d states of Pd.

The existence of the complex Pd–Ge polyanion and the electronegativity difference between Pd and Ge could be invoked to explain the trend of Ge species charges, neglecting at a first approximation Lu atoms (see Table 6.4 and Fig. 4.2). The Ge dumbbell in $Lu_3Pd_4Ge_4$ has four neighboring Pd atoms, instead those in $Lu_5Pd_4Ge_8$

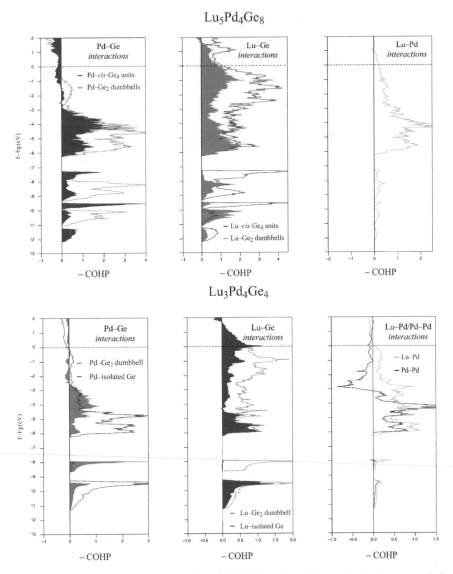

Fig. 6.8 Crystal Orbital Hamilton Populations (–COHP) for selected interactions for the two studied compounds. From [9]

install six Pd–Ge polar interactions. As a result, the latter Ge species has lower negative charges. The same is true for isolated (0*b*)Ge atoms with six palladium atoms around in Lu₃Pd₄Ge₄: its charge approaches to zero. Within crystal structure, the number of Pd–Ge contacts is the same for terminal and central atoms of *cis*-Ge₄ units; thus, their charges trend is similar as for ideal *cis*-Ge₄$^{10-}$ anion, terminal atoms being more negative.

Inside $Lu_3Pd_4Ge_4$ the presence of a Pd–Pd short distance (2.75 Å) can be highlighted. The corresponding—COHP plots are similar to those reported for Ca_2Pd_3Ge [42] showing a sharp antibonding character around −3 eV commonly attributed to enhanced repulsion between filled d states of Pd. Nevertheless, they are of bonding type in average as deducible from the—iCOHP values for this interaction (0.97 eV/cell), comparable to those reported in [42].

The remaining Lu–Pd and Ge–Lu interactions are weaker being however very similar for both germanides. All of them are of bonding type; the Lu–Pd interactions are practically optimized at the Fermi level whereas the Ge–Lu show many unoccupied bonding states well above E_F. The latter feature was also reported for other binary (Ca_5Ge_3 [39] and CaSi [43]) and ternary (La_2ZnGe_6 [44]) tetrelides, being related to a mixing between the d states of Ca/La with the p of Ge. Since for La_2ZnGe_6, complete chemical bonding study (see Chap. 4) was also performed during the doctorate on the basis of the real space techniques, the presence of multiatomic Ge–La bonding was confirmed and also revealed in La_2PdGe_6 and Y_2PdGe_6 analogues. Thus, Ge–Lu interactions can be described as polar covalent also for the title compounds.

As stated above, the weak Lu–Pd interaction are always of bonding type, confirming again the presence of R–Pd heteropolar bonds similarly to Pd-containing 2:1:6 stoichiometry phases (and also to the other T-containing ones, with $T = $ Zn, Cu, Ag).

These findings are also in very good agreement with the first analyses of the ELI-D for $Lu_5Pd_4Ge_8$ and $Lu_3Pd_4Ge_4$. In Fig. 6.9 the ELI-D planar distribution is highlighted for Ge_2 (b) dumbbells and cis-Ge_4 (a) fragments, confirming the covalent

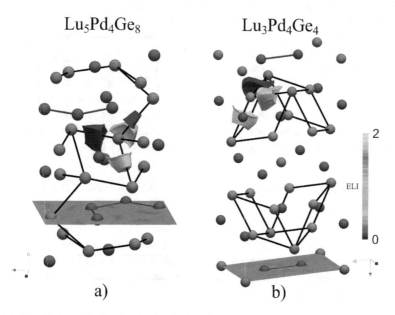

Fig. 6.9 ELI-D planar distribution for the cis-Ge_4 chain (**a**) and the Ge_2 dumbbell (**b**) and ELI-D basins for Lu–Pd interactions within $Lu_5Pd_4Ge_8$ (**a**) and $Lu_3Pd_4Ge_4$ (**b**) unit cells. (Ge: orange; Lu: green; Pd: grey)

nature of Ge–Ge interactions. In the same plot, the ELI-D basins for the interesting Lu–Pd bonds are also shown.

Further position-space chemical bonding analysis for these phases are planned, including also the investigation of the fine structure of the ELI-D based on the topology of its (relative) Laplacian.

6.1.7 Conclusions on Lu$_5$Pd$_4$Ge$_8$ and Lu$_3$Pd$_4$Ge$_4$ Intermetallics

The two new Lu$_5$Pd$_4$Ge$_8$ and Lu$_3$Pd$_4$Ge$_4$ polar intermetallics were synthesized and characterized. They were found to crystallize in the mP34–Tm$_5$Pd$_4$Ge$_8$ and oI22–Gd$_3$Cu$_4$Ge$_4$ structures respectively. A detailed description of crystal structure solution in the case of the non-merohedral twinned crystal of Lu$_5$Pd$_4$Ge$_8$ was proposed, highlighting the difficulties/problems encountered here along with practical suggestions to manage them.

Joined crystal structure analysis and DOS/COHP based chemical bonding studies confirmed that the assumption of a complete formal charge transfer from metals to Ge is a too strict approximation. The presence of covalently bonded Ge atoms (Ge$_2$ dumbbells and cis-Ge$_4$ fragments) was confirmed.

When compared with the corresponding Ge$_2{}^{6-}$ and Ge$_4{}^{10-}$ Zintl anions, they both show a reduced antibonding states population, caused by other bonding interactions with the surrounding Pd and Lu atoms. In fact, COHP curves and iCOHP values for Ge–Pd and Ge–Lu interactions confirm it, with the latter being of bonding type until E_F and above. This feature, also observed in other similar compounds (including the here studied La$_2$ZnGe$_6$ phase), confirm the relevance of Ge–R bonding in polar ternary germanides. The same is also true for the Lu–Pd interaction. Moreover, a short Pd–Pd distance (2.75 Å) was found in Lu$_3$Pd$_4$Ge$_4$ and interpreted as a weak metal–metal bond. The nature of chemical bonding within title compounds has been targeted for future analysis on the basis of the position-space methods, including the ELI-D fine structure which was proven to be particularly useful to interpret Ge–R and R–T interactions. Preliminary ELI-D based results are encouraging.

6.2 Synthesis and Crystal Structure for the New Yb$_2$PdGe$_3$ Germanides

Preliminary results obtained on the new AlB$_2$ derivative, the Yb$_2$PdGe$_3$ intermetallic, are presented in the following paragraphs. Details on its preparation both from metal flux (Indium) and direct synthesis, crystal structure solution and group-subgroup relations are presented.

6.2.1   Synthesis and Phase Analysis

As already mentioned in Chap. 5, a metal flux synthesis, employing Indium as solvent, was performed applying the following thermal cycle:

$$25\,°C \rightarrow (10\,°C/min) \rightarrow 750\,°C(24h) \rightarrow (-0.5\,°C/min) \rightarrow 25\,°C.$$

Stoichiometric amounts of Yb, Pd and Ge, giving the nominal composition $Yb_{21}Pd_7Ge_{72}$, were put in an arc-sealed Ta crucible with a 1:45 molar excess of indium, to obtain a total mass of about 3 g. The main purpose of this synthesis was the preparation of $Yb_2Pd_3Ge_5$ single crystal. Even though no traces of the wanted phases were detected, another new compounds was revealed. Then, it was decided to isolate it from the In flux with a centrifugation at 300 °C [T_m(In) = 156.6 °C]. Whereupon, the obtained powders were sonicated within glacial acetic acid in order to selectively oxidize the residual In.

At this point, since it is well known that the use of a metal flux could stabilize metastable phases [45] (see also Chap. 4), the existence of the new Yb-containing germanide was tested also by direct synthesis in resistance furnace. The stoichiometric amount of the constituents, giving the $Yb_{33.3}Pd_{16.7}Ge_{50.0}$ nominal composition, were put in an arc-sealed Ta crucible obtaining a total mass of about 0.7 g. The crucible was then closed in an evacuated quartz phial to prevent its oxidation at the employed temperature and put in a resistance furnace were the following thermal cycle was applied:

$$25\,°C \rightarrow \left(10\,°C/min\right) \rightarrow 950\,°C(1h) \rightarrow \left(-0.2\,°C/min\right) \rightarrow$$
$$350\,°C \rightarrow (\text{furnace switched off}) \rightarrow 25\,°C.$$

A continuous rotation, at a speed of 100 rpm, was applied to the phial during the thermal cycle.

6.2.2   Samples Characterization by SEM/EDXS Analysis and XRPD

Results of SEM/EDXS and XRPD characterization reveal the presence of the Yb_2PdGe_3 phase in both samples prepared by metal flux and direct synthesis (see Table 6.8).

The metal flux synthesized, under the applied synthetic conditions, turned out to be reactive forming big (~500μm) Pd_3In_7 single crystals (Fig. 6.10a). Anyway, good quality crystals of Yb_2PdGe_6, recrystallized Ge and the new Yb_2PdGe_3 were detected both by SEM-EDXS (see Fig. 6.10a and c) and X-ray powder pattern. The latter was measured on powders obtained after the separation procedure. It is worth to mention that the measured composition for Yb_2PdGe_3 agrees with that reported by

Table. 6.8 Structural data (XRPD) and elemental atomic percent composition (SEM/EDXS) for Yb–Pd–Ge samples prepared by metal flux and direct synthesis in resistance furnace

Synthetic method	Phases	Composition by (EDXS) [at %]				Pearson symbol prototype	Lattice parameters [Å]		
		Yb	Pd	Ge	In		a	b	c
In flux[a]	**Yb_2PdGe_3**	**34.2**	**17.6**	**48.2**	–	**$hP24$-Lu_2CoGa_3**	**8.474(2)**	**8.474(2)**	**8.147(3)**
	Yb_2PdGe_6	23.0	12.1	64.9	–	$oS72$-$Ce_2(Ga_{0.1}Ge_{0.7})_9$	8.150(2)	7.990(1)	21.847(5)
	Pd_3In_7	–	31.4	–	68.6	$cI40$-Ru_3Sn_7	9.4275(6)	9.4275(6)	9.4275(6)
	Ge	–	–	100.0	–	$cF8$-C	5.654		
Direct synthesis	**Yb_2PdGe_3**	**33.7**	**18.4**	**47.9**	–	**$hP24$-Lu_2CoGa_3**	**8.464(1)**	**8.464(1)**	**8.147(2)**
	Yb_2PdGe_6[b]	23.2	12.2	64.6					

[a]Sample from which Yb_2PdGe_3 single crystal was taken
[b]Phases identified only by EDXS analysis

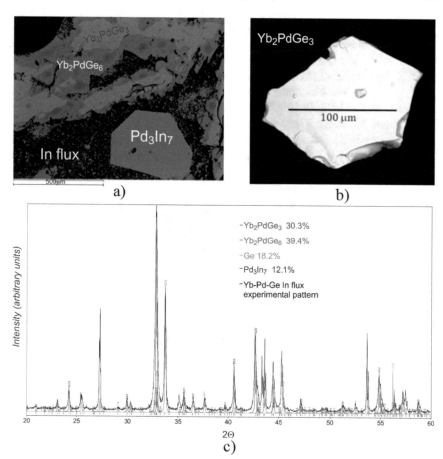

Fig. 6.10 BSE micrographs and X-ray powder patterns for Yb–Pd–Ge sample synthesized in metal flux prior (**a**) and after (**b–c**) the In removal by means of both centrifugation and selective oxidation

Seropegin et al. [46] for the phase Yb(Pd,Ge)$_2$. For more details on their structures, see Paragraph 6.3.3. The flux-synthesized sample was enclosed in an Ar-filled quartz phial as reported in Chapter 5, Fig. 5.4. After heating it for one minute at 300 °C (well above the T_m(In)) within a resistance furnace, it was quickly removed and put into a centrifuge at a speed of 600 rpm for about 1 min. This procedure was repeated five times aiming to ensure a quantitative flux separation. The residual In was selectively oxidized by immersion and sonication of the crystals in glacial acetic acid [CH$_3$COOH$_{(l)}$] for about 2 h. The peaks associated to cubic In were not revealed in the X-ray powder patter (Fig. 6.10c). Good quality Yb$_2$PdGe$_3$ crystals, showing a plate-like morphology, were detected with SEM/EDXS (Fig. 6.10b). It is worth to note that the crystal shown in Fig. 6.10b is the one selected and analysed by means of X-ray single diffraction.

The existence of the Yb$_2$PdGe$_3$ phase was also checked by direct synthesis in resistance furnace. Even though samples synthesized by slow cooling in the resistance furnace are usually characterized by an inhomogeneous microstructure, containing several different phases, this sample was revealed to be almost Yb$_2$PdGe$_3$ single phase (see Fig. 6.11). A small amount of Yb$_2$PdGe$_6$ and another phase (bright striping) was detected with SEM/EDXS (Fig. 6.11a) together with some unindexed low intensity peaks around 35° in the X-ray powder pattern (Fig. 6.11b).

Further synthesis and annealing are planned in order to improve the yield of the Yb$_2$PdGe$_3$ aiming to perform physical properties measurement, *i.e.* to investigate the valence of Yb.

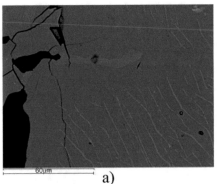

a)

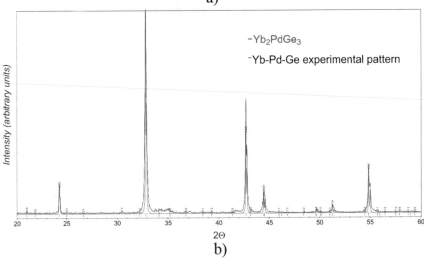

b)

Fig. 6.11 (**a**) SEM micrograph (BSE mode) and (**b**) X-ray powder patter for the sample of Yb$_{33.3}$Pd$_{16.7}$Ge$_{50.0}$ nominal composition. **a** The grey phase is Yb$_2$PdGe$_3$ and the bright is Yb$_2$PdGe$_6$. The striping-like morphology phase composition was not measured due to its fine structure

6.2.3 Crystal Structure of Yb₂PdGe₃ as a 2a × 2a × 2c AlB₂ Superstructure

It's known from the literature on the existence of Yb(Pd,Ge)$_2$ ternary compound whose crystal structure was interpreted as AlB$_2$-like with unit cell parameters $a = 4.2276$, $c = 4.0686$ Å [46]. The Pd content for this phase was roughly estimated as 16.7 at%. Pd and Ge occupy statistically the $2d$ crystallographic site, forming planar hexagonal graphite like layers.

In the current investigation a single crystal (Fig. 6.10b) of this phase was extracted from the metal flux medium (In) after centrifugation and selective oxidation. Since the composition of a set of selected crystals, preliminary checked by SEM/EDXS, agree with those reported by Seropegin [46] the correctness of the AlB$_2$-like structure type was checked.

The indexation of the most intense peaks collected for title crystal gives hexagonal unit cell with. $a \sim 4.23$, $c \sim 4.07$ Å. Successively the data set was integrated and appropriately corrected before merging intensities. The structural model was quickly found by charge flipping algorithm implemented in JANA2006 [27]. It was confirmed to be AlB$_2$-like ($hP3$, $P6/mmm$) with residuals R1/wR2 close to 10%. Anyway, it was decided to consider small intensity super reflections in crystal structure determination. In this case, the indexation was straightforward, giving a unit cell eight times bigger with respect to the previous one (parent type). The crystal symmetry remains always hexagonal with $a \sim 8.47$, $c \sim 8.15$ Å. In the direct space the unit cell of supercell is related to the parent type by the following transformation expressed as:

$$A = \begin{pmatrix} 0 & 2 & 0 \\ 2 & 0 & 0 \\ 0 & 0 & -2 \end{pmatrix}$$

In the reconstructed $hk0$ precession image (Fig. 6.12) is also shown the reciprocal space relation between parent and derivative unit cells. Moreover, it is evidenced the intensity difference between principal and super reflections. One may note the ordered distribution of weak intensities peaks and only a very small fractions of spurious peaks.

The analysis of the systematic absences suggests a primitive lattice centering and numerous possible space groups ($P6/mcc$, $P6cc$, $P6_3/mcm$, $P-6c2$, $P6_3cm$, $P6_3/mmc$, etc....). Moreover, the $|E^2-1|$ criteria was ~ 1.6, being noticeably far from the ideal value of 1 (centrosymmetric space group). These observations can be reasonably explained by the big number of weak super reflections present in the data set. The search for the most chemically reasonable structure models was started from the highest symmetry space groups. In this way a promising structural model of Yb₂PdGe₃ composition was found in $P6_3/mmc$ space group.

The further structure refinements were carried out by full-matrix least-squares methods on $|F^2|$ using the SHELX programs [47]. The site occupancy factors of

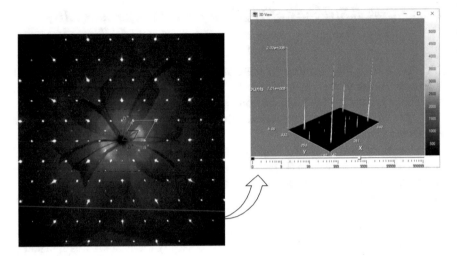

Fig. 6.12 Observed intensity profiles for $hk0$ zone demonstrate the relation between the unit cells of AlB$_2$ parent type (green lines) and that of the superstructure (purple lines). The presence of weak super-reflections is well visible and also shown in the 3D mode to the right

Table. 6.9 Crystallographic data for Yb$_2$PdGe$_3$ single crystal taken from the sample prepared by flux-synthesis (space group: $P6_3/m\,2/m\,2/c$, № 194; Pearson's symbol-prototype: $hP24$- Lu$_2$CoGa$_3$; Z=4)

Empirical formula	Yb$_2$PdGe$_3$
EDXS data	Yb$_{34.2}$Pd$_{17.6}$Ge$_{48.2}$
M$_w$, [g/mol]	670.25
a [Å]	8.466(2)
b [Å]	8.466(2)
c [Å]	8.147(8)
V [Å^3]	505.8(2)
Calc. density [g/cm^3]	8.802
abs coeff (μ), mm^{-1}	57.5
Unique reflections	355
Reflections I > 2σ(I)	221
R$_{sigma}$	0.0179
Data/parameters	221/50
GOF on F^2 (S)	1.421
R indices [I > 2σ(I)]	R$_1$ = 0.0209; wR$_2$ = 0.0881
R indices [all data]	R$_1$ = 0.0315; wR$_2$ = 0.1014
$\Delta\rho_{fin}$ (max/min), [e/Å^3]	1.71/ −1.69

Table. 6.10 Atomic coordinates and equivalent isotropic displacement parameters (Å^2) for the investigated Yb_2PdGe_3 single crystal

Atomic param	Yb1 (6h)	Yb2 (2b)	Pd (4f)	Ge (12 k)
x/a	0.50022(3)	0	1/3	0.16613(2)
y/b	0.00044(6)	0	2/3	0.33226(4)
z/c	0.25	0.25	0.5008(1)	0.00008(8)
U_{eq} (Å^2)	0.0090(2)	0.0073(2)	0.0085(3)	0.0096(3)

all species were checked for deficiency, in separate cycles of refinement, obtaining values very close to unity. At this point neither deficiency nor statistical mixture were considered and stoichiometric Yb_2PdGe_3 models were further anisotropically refined giving acceptable residuals and flat difference Fourier maps. Selected crystallographic data and structure refinement parameters for these crystals are listed in Tables 6.9 and 6.10.

Then, the $Yb(Pd,Ge)_2$ model with Pd/Ge mixing is revised with an ordered one that could be viewed as a derivative of the AlB_2 structure, as already described in

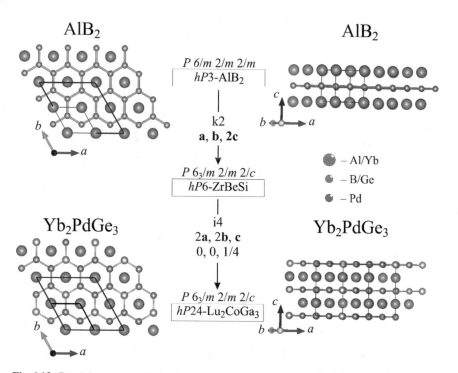

Fig. 6.13 Bärnighausen tree relating the AlB_2 aristotype and its Yb_2PdGe_3 derivative. The graphite like layers composed by B_6 (AlB_2) and Ge_6/Ge_4Pd_2 (Yb_2PdGe_3) are evidenced within the unit cells. The doubling of a/b axis (left) and c axis (right) is also evident. The black and blue unit cells (left) highlit the metric relations between AlB_2 and Yb_2PdGe_3

[48]. The structural hierarchy tree was reconstructed for the branch of interest and reported in Fig. 6.13.

The structure plot of both the title compound (Yb_2PdGe_3) and the aristotype (AlB_2) are also reported in order to point out the metrical relations between the involved structures. Starting from the AlB_2 type, symmetry is reduced by a *klassengleiche* symmetry reduction of index 2 ($k2$), leading to the $P6_3/m\ 2/m\ 2/c$ space group, with a doubling of the c axis (ZrBeSi-like structure). An additional *isomorphic* transition of index 4 ($i4$) yields to the Yb_2PdGe_3 structure, with the doubling also of a and b axis which results in an unit cell eight times bigger than that of AlB_2. Yb-atoms occupy two different positions: $2b$ and $6\ h$. Both of them are placed at the centre of hexagons: the former composed of Ge_6 and the latter of Ge_4Pd_2 (with Pd species placed in *para*). On the basis of Ge–Ge (2.44 Å) and Ge–Pd (2.45 Å) interatomic distances it is reasonable to interpret the puckered layers as composed by covalently bonded hexagons of Ge and Pd species. Thus, the study of chemical bonding for Yb_2PdGe_3 would be of great interest in the framework of what was called by Miller et al. the chemistry of inorganic "graphene" [49, 50].

6.2.4 Conclusions on the Yb_2PdGe_3 Germanide

The new Yb_2PdGe_3 was prepared by both metal flux synthesis, performed with Indium, and direct synthesis in resistance furnace. A good quality single crystal was extracted from the flux-synthesized samples. Its structure turned out to be $hP24$-Lu_2CoGa_3, which is an ordered derivative of AlB_2. In fact, it can be described as composed of puckered layers composed by Ge_6 and Ge_4Pd_2 hexagons, with Ge–Ge and Ge–Pd distances being almost identical (2.44–2.45 Å). Obtained results suggest to investigate both physical properties and chemical bonding for Yb_2PdGe_3 and also the existence of this phase with other R metals since the $R(Pd,Ge)_2$ compounds are often reported with the AlB_2 type structure and could be revised as ordered derivatives.

References

1. Freccero R, Solokha P, Proserpio DM, Saccone A, De Negri S (2017) Dalton Trans 46:14021–14033
2. Solokha P, Freccero R, De Negri S, Proserpio DM, Saccone A (2016) Struct Chem 27:1693–1701
3. Freccero R, Choi SH, Solokha P, De Negri S, Takeuchi T, Hirai S, Mele P, Saccone A (2019) J Alloys Compd 783:601–607
4. Villars P, Cenzual K (2018/19) Pearson's crystal data. ASM International, Ohio, USA
5. Niepmann D, Prots YM, Pöttgen R, Jeitschko W (2000) J. Solid State Chem 154:329–337
6. Heying B, Rodewald UC, Pöttgen R (2017) Z Kristallogr 232:435–440

7. Rodewald UC, Heying B, Hoffmann RD, Niepmann D (2009) Z Naturforsch 64b:595–602
8. Chabot B, Parthè E, Less-Comm J (1985) Metals 106:53–59
9. Freccero R, Solokha P, Proserpio DM, Saccone A, De Negri S (2018) Crystals 8:205
10. Krie r G, Jepsen O, Burkhardt A, Andersen OK (2000) The TB-LMTO-ASA program; Version 4.7; Max-Planck-Institut für Festkörperforschung, Stuttgart, Germany
11. Barth U, Hedin LA (1972) Local Exchange-Correlation Potential for the Spin Polarized Case: I. J. Phys. Chem. C 5:1629–1642
12. Steinberg S, Dronskowski R (2018) Crystals 8:225. https://doi.org/10.3390/cryst8050225
13. Eck B (199–2018) Wxdragon. Aachen, Germany. Available at https://wxdragon.de
14. Bader RFW (1990) Atoms in molecules: a quantum theory. Oxford University Press, Oxford
15. Kohout M (2011) Program *Dgrid,* version 4.6, Radebeul
16. Perdew JP, Wagn Y (1992) Phys Rev B 45:13244
17. Giannozzi P, Baroni S, Bonini N, Calandra M, Car R, Cavazzoni C, Ceresoli D, Chiarotti GL, Cococcioni M, Dabo I, Dal Corso A, de Gironcoli S, Fabris S, Fratesi G, Gebauer R, Gerstmann U, Gougoussis C, Kokalj A, Lazzeri M, Martin-Samos L, Marzari N, Mauri F, Mazzarello R, Paolini S, Pasquarello A, Paulatto L, Sbraccia C, Scandolo S, Sclauzero G, Seitsonen AP, Smogunov A, Umari P, Wentzcovitch RM (2009) J Phys Condens Matter 21:395502
18. Mealli C, Proserpio DM (1990) MO Theory made visible. J Chem Educ 67:399–403
19. Mealli C, Ienco A, Proserpio DM (1998) Book of abstracts of the XXXIII ICCC. Florence, 510
20. Lowe J, Peterson K (2005) Quantum chemistry, 3rd edn. Academic Press
21. Hoffmann R (1988) Solids and surfaces: a chemist's view of bonding in extended structures. VCH, New York
22. Solokha P, De Negri S, Saccone A, Proserpio DM (2012) On a non-merohedrally twinned Tb3Ge5 crystal. In: XII international conference on crystal chemistry of intermetallic compounds, Lviv (Ukraine), 22–26 September, 120
23. Budnyk S, Weitzer F, Kubata C, Prots Y, Akselrud LG, Schnelle W, Hiebl K, Nesper R, Wagner FR, Grin Y (2006) J Solid State Chem 179:2329
24. Shcherban O, Savysyuk I, Semuso N, Gladyshevskii R, Cenzual K (2009) Chem Alloy Met 2:115
25. Müller P (2006) Crystal structure refinement: a crystallographer's guide to SHELXL. Oxford University Press, Oxford, UK
26. Clegg W (2009) Crystal structure analysis: principle and practice, 2nd edn. Oxford University Press, Oxford, UK
27. Petricek V, Dusek M, Palatinus L (2014) Z Kristallogr 229(5):345–352
28. Bruker (2014) APEX2, SAINT-Plus, XPREP, SADABS, CELL_NOW and TWINABS. Bruker AXS Inc., Madison, Wisconsin, USA
29. Cooper RI, Gould RO, Parsons S, Watkin DJ (2002) J Appl Crystallogr 35:168–174
30. Farrugia LG (2012) WinGX and ORTEP for Windows: an update. J Appl Cryst 45:849–854
31. Gelato LM, Parthé E (1987) J Appl Crystallogr 20:139
32. Parthé E, Gelato L, Chabot B, Penzo M, Cenzual K, Gladyshevskii R (1993) TYPIX. standardized data and crystal chemical characterization of inorganic structure types, vol 1. Heidelberg: Springer-Verlag, 260 p
33. Yu Prots R, Demchyna U, Burkhardt US (2007) Z Kristallogr 222:513–520
34. H. Wondratschek, U. Müller (2004), International tables for crystallography, vol. A1, symmetry relations between space groups. Kluwer Academic Publishers, Dordrecht
35. Müller U (2013) Symmetry relationships between crystal structures. Applications of crystallographic group theory in crystal chemistry, Oxford University Press, Oxford, UK
36. Pöttgen R, Anorg Z (2014) Allg Chem 640:869–891
37. Landrum GA, Dronskowski R (2000) Angew Chem Int Ed 39:1560–1585
38. Lin Q, Corbett JD (2009) Inorg Chem 48:5403–5411
39. Mudring AV, Corbett J (2004) J Am Chem Soc 126:5277–5281
40. Siggelkow L, Hlukhyy V, Fässler TF (2012) J. Solid State Chem 191:76–89
41. Eisenmann B, Schäfer H, Anorg Z (1974) Allg Chem 403:163–172

42. Doverbratt I, Ponou S, Lidin S (2013) J Solid State Chem 197:312–316
43. Kurylyshyn IM, Fässler TF, Fischer A, Hauf C, Eickerling G, Presnitz M, Scherer W (2014) Angew Chem Int Ed 53:3029–3032
44. Solokha P, De Negri S, Proserpio DM, Blatov VA, Saccone A (2015) Inorg Chem 54:2411. https://doi.org/10.1021/ic5030313
45. Kanatzidis MG, Pöttgen R, Jeitschko W (2005) Angew Chem Int Ed 44:6996
46. Yu D, Seropegin OL, Borisenko OI, Bodak VN, Nikiforov MV, Kovachikova Y, Kochetkov V (1994) J Alloys Compd 216:259–263
47. Sheldrick GM (2008) Acta Cryst A 64:112–122
48. Hoffmann RD (2001) R. Pöttgen. Z Kristallogr 216:127–145
49. Miller GJ, Schmidt MW, Wang F, You T-S (2011), Quantitative advances in the Zintl–Klemm Formalism. In: Fässler TF (ed) Zintl phases, principles and recent developments. Springer Series in Structure and Bonding 139, Springer Science+Business Media Dordrecht
50. You TS, Miller MJ (2009) Inorg Chem 48:6391–6401

Chapter 7
The R_4MGe_{10-x} (R = Rare Earth Metal; M = Li, Mg) Compounds: A New Family of Ternary Germanides

The new ternary germanide La_4MgGe_{10-x} was synthesized by De Negri et al. [1] during the investigation of the phase equilibria at 500 °C for the La–Mg–Ge system (see Chap. 1). During the study on the R_2LiGe_6 (see Chap. 4, Table 4.3) compounds, the 4:1:(10–x) stoichiometry phase turned out to form also with Li and R = La, Ce, Pr, Nd. These findings fostered the investigation of the title compounds existence along the R series with both Mg and Li. In this chapter preliminary results on their crystal structure solution are also presented.

7.1 Synthesis

Samples of about 0.8 g with $R_{26.6}M_{6.7}Ge_{66.7}$ (R = Y; La–Nd; Sm–Lu; M = Li, Mg) nominal compositions, corresponding to the R_4MGe_{10} stoichiometry, were prepared by direct synthesis in resistance furnace. Stoichiometric amounts of components were placed in an arc–sealed Ta crucible in order to prevent oxidation; the constituents of the Li–containing alloys and the $Eu_{26.6}Mg_{6.7}Ge_{66.7}$ sample were weighed and sealed into the crucible inside an inert atmosphere glove box. The Ta crucible was then closed in an evacuated quartz phial to prevent oxidation at high temperature, and finally placed in a resistance furnace, where the following thermal cycle was applied:

$$25\,°C \rightarrow (10\,°C/min) \rightarrow 950\,°C\ (1\ h) \rightarrow (-0.2\,°C/min)$$
$$\rightarrow 350\,°C \rightarrow \text{furnace switched off}$$

A continuous rotation, at a speed of 100 rpm, was applied to the phial during the thermal cycle. These synthetic conditions were chosen with the aim to obtain samples containing crystals of good quality and size, suitable for further structural studies. All the synthesized alloys are very brittle; the Mg–containing ones are also stable in air, contrary to the Li–containing and the $Eu_{26.6}Mg_{6.7}Ge_{66.7}$. $TaGe_2$ intermetallic was often detected in Li alloys, indicating a crucible contamination of some samples.

Microstructure examination as well as qualitative and quantitative analyses were performed by SEM–EDXS using cobalt standard for calibration. The problem to get reliable EDXS composition when dealing with compounds containing both Mg and Ge, was already reported in [1]. In fact, R–Mg–Ge samples are always affected by a systematic error due to the energy resolution limit of the spectrometer, which leads to a severe peak overlap between the only line of Mg (K) and some Ge L. As a consequence, the magnesium/germanium concentration ratio is overestimated; in addition, the magnitude of the error depends on the composition itself without a regular trend. In average, the Mg concentration provided by the software is about 3 to 7 at. % higher than the real value when measuring inside the grains of single phases, even higher when measuring the overall alloy compositions. The R content is generally reliable. For the La–Mg–Ge alloys [1] no significant improvements were obtained by adopting the procedures suggested by the INCA Energy Operator Manual for similar cases, including accurate quant optimization, different quant configurations and profile optimization.

On the other hand, EDXS probe does not reveal Li. Thus, binary Li intermetallics are detected as pure elements, ternary as binary, etc.… For example, the wanted phase, was detected as a ~ "R_2Ge_5" binary phase. Hence, taking into account the previous considerations, the EDXS data recorded on all samples were simply used as guidelines to identify phases, whose exact composition was normally derived from the crystal structure.

Crystals of Ce_4MgGe_{10-x}, Nd_4MgGe_{10-x} and Nd_4LiGe_{10-x} compounds were extracted from mechanically fragmented alloys and analysed through X–ray single crystal diffraction.

7.2   On the Existence of R_4MgGe_{10-x} and R_4LiGe_{10-x} Compounds

Results of SEM/EDXS and XRPD characterization for all the synthesized alloys are listed in Tables 7.1 and 7.2.

The measured EDXS compositions and the XRPD pattern confirm the formation of the 4:1:(10–x) phase in samples with R = La–Nd, Sm, Gd–Dy, showing the same trend with Li and Mg. In the Yb–Li–Ge alloy, a phase with $Yb_{34.4}Ge_{68.6}$ composition was detected; due to its low amount, it was not possible to assign the correct structure so that one more representative would be added to the investigated series. It should be noted that the R–Li–Ge alloys are prone to react with Ta crucibles, contrary to

Table 7.1 Structure data and atomic per cent composition of the constituting elements for $R_{26.6}Mg_{6.7}Ge_{66.7}$ samples. R_4MgGe_{10-x} phases are listed in bold

R	Phases	Composition by (EDXS) [at.%]			Pearson symbol prototype	Lattice parameters [Å]		
		R	Mg	Ge		a	b	c
Y	YGe$_2$	32.0	9.2	58.8	$tI12$–ThSi$_2$	4.0635(8)		13.669(5)
	Ge		9.0	91.0	$cF8$–C	5.6534(5)		
	Mg$_2$Ge		68.4	31.6	$cF12$–CaF$_2$	6.3812(8)		
	Y$_3$Ge$_5$[a]				$oF64$– Y$_3$Ge$_5$	5.740(2)	17.320(9)	13.656(3)
Ce	**Ce$_4$MgGe$_{10-x}$**	**25.3**	**13.7**	**61.0**	**$mS60$-y–La$_4$MgGe$_{10-x}$**	**8.745(4)**	**8.567(4)** **β = 97.15(8)**	**17.499(5)**
	Ge		10.2	89.8	$cF8$–C	5.649(2)		
	Mg$_2$Ge		31.0	69.0	$cF12$–CaF$_2$	6.364(2)		
Pr	**Pr$_4$MgGe$_{10-x}$**	**24.9**	**13.1**	**62.0**	**$mS60$-y–La$_4$MgGe$_{10-x}$**	**8.656(3)**	**8.370(2)** **β = 97.11(6)**	**17.620(6)**
	Ge[a]				$cF8$–C	5.657(1)		
	Mg$_2$Ge		60.6	41.4	$cF12$–CaF$_2$	6.347(2)		
	PrGe$_2$	32.5	7.3	60.3	$tI12$ – ThSi$_2$	o		
Nd	**Nd$_4$MgGe$_{10-x}$**	**24.5**	**12.1**	**63.4**	**$mS60$-y–La$_4$MgGe$_{10-x}$**	**8.530(6)**	**8.256(4)** **β = 97.00(4)**	**17.571(5)**
	Ge		5.8	94.2	$cF8$–C	5.655(1)		
	Mg$_2$Ge		62.5	37.5	$cF12$–CaF$_2$	6.364(2)		
	NdGe$_{1.667}$	31.5	9.7	58.7	$tI12$–1.34–ThSi$_{2-x}$	4.230(1)		13.906(5)
Sm	**Sm$_4$MgGe$_{10-x}$**	**25.1**	**12.2**	**62.7**	**$mS60$-y–La$_4$MgGe$_{10-x}$**	**8.400(4)**	**8.140(1)** **β = 97.97(3)**	**17.397(4)**
	Ge		6.8	93.1	$cF8$–C	5.648(1)		

(continued)

Table 7.1 (continued)

R	Phases	Composition by (EDXS) [at.%]			Pearson symbol prototype	Lattice parameters [Å]		
		R	Mg	Ge		a	b	c
	Mg$_2$Ge		53.7	46.3	$cF12$–CaF$_2$	6.361(2)		
	SmGe$_{2-x}$	32.1	6.7	61.2	$tI12$–ThSi$_{2-x}$	4.170(1)		13.766(7)
Eu	Ge		9.6	90.4	$cF8$–C	5.656(1)		
	Mg$_2$Ge		69.2	30.7	$cF12$–CaF$_2$	6.368(5)		
	EuGe$_2$	28.7	9.7	61.6	$hP3$–EuGe$_2$	4.095(1)		4.998(2)
Gd	**Gd$_4$MgGe$_{10-x}$**	**26.0**	**11.6**	**62.5**	**$mS60$–y–La$_4$MgGe$_{10-x}$**	**8.300(5)**	**8.075(3)** **β = 96.61(5)**	**17.361(7)**
	Ge			100	$cF8$–C	5.652(2)		
	Mg$_2$Ge		68.8	31.2	$cF12$–CaF$_2$	6.367(4)		
	GdGe$_{2-x}$	36.7		63.3	$tI12$–1.36–ThSi$_{2-x}$	4.127(1)		13.726(4)
Tb	**Tb$_4$MgGe$_{10-x}$**[a]				**$mS60$–y–La$_4$MgGe$_{10-x}$**	**8.224(4)**	**8.019(12)** **β = 96.93(7)**	**17.284(11)**
	Ge		7.1	92.6	$cF8$–C	5.654(1)		
	Mg$_2$Ge		68.1	31.9	$cF12$–CaF$_2$	6.379(1)		
	TbGe$_2$	32.6		67.4	$oS24$–TbGe$_2$	4.107(1)	29.832(8)	3.984(1)
	Tb$_3$Ge$_5$	37.1		62.9	$oF64$–Y$_3$Ge$_5$	5.765(2)	17.242(4)	13.668(1)
Dy	**Dy$_4$MgGe$_{10-x}$**	**26.4**	**10.8**	**62.8**	**$mS60$–y–La$_4$MgGe$_{10-x}$**	**8.194(5)**	**7.978(5)** **β = 96.97(9)**	**17.297(25)**
	Ge[a]				$cF8$–C	5.648(2)		
	Mg$_2$Ge		67.7	32.3	$cF12$–CaF$_2$	6.383(4)		

(continued)

Table 7.1 (continued)

R	Phases	Composition by (EDXS) [at.%]			Pearson symbol prototype	Lattice parameters [Å]		
		R	Mg	Ge		a	b	c
	DyGe$_{1.75}$	66.4		33.6	$oS24$–2–DyGe$_{2-x}$	4.075(1)	29.87(2)	3.968(1)
	DyGe$_3$	26.5		74.5	$oS16$–CeNi$_{(1-x)}$Si$_2$	4.047(4)	20.722(17)	3.904(3)
Ho	HoGe$_{1.85}$	34.1		65.9	$oS28$–5.22–YGe$_{1.82}$	4.072(2)	29.617(7)	3.9020(9)
	HoGe$_3$	25.2		74.8	$oS16$–CeNi$_{(1-x)}$Si$_2$	4.007(1)	20.64(1)	3.884(4)
	Ge^a				$cF8$–C	5.651(1)		
	Mg$_2$Ge		67.0	33.0	$cF12$–CaF$_2$	6.361(2)		
Er	Er$_2$Ge$_5$	29.4		70.6	$oP14$–Er$_2$Ge$_5$	3.888(3)	4.010(2)	18.171(9)
	Ge			100	$cF8$–C	5.651(2)		
	Mg$_2$Ge		68.2	31.8	$cF12$–CaF$_2$	6.360(3)		
	ErGe$_3$	26.0		73.9	$oS16$–CeNi$_{(1-x)}$Si$_2$	4.011(9)	20.54(5)	3.881(3)
	ErGe$_{1.835}$	35.5		64.5	$oS24$–1.32–DyGe$_{2-x}$	4.046(3)	29.70(8)	3.892(7)
Yb	Yb$_3$Ge$_8$	25.6	9.3	65.1	$aP22$–Yb$_3$Ge$_8$	7.286(7) $\alpha=104.3(1)$	7.385(8) $\beta=102.9(1)$	10.35(2) $\gamma=110.0(1)$
	Ge		8.3	91.7	$cF8$–C	5.667(1)		
	Mg$_2$Ge		66.0	34.0	$cF12$–CaF$_2$	6.3564(7)		
	Yb$_3$Ge$_5$	35.7	8.3	56.0	$hP8$–Th$_3$Pd$_5$	6.842(8)		4.167(4)

^{a} Phases not detected with SEM/EDXS

Table 7.2 Structure data and atomic per cent composition of the constituting elements for $R_{26.6}$Li$_{16.7}$Ge$_{66.7}$ samples. R_4LiGe$_{10-x}$ phases are listed in bold

R	Phases	Composition by (EDXS) [at.%]		Pearson symbol prototype	Lattice parameters [Å]		
		R	Ge		a	b	c
Y	YGe$_{2-x}$	32.8	67.2	$oS24$–TbGe$_2$	4.134(9)	29.98(3)	3.959(9)
	Ge	–	100	$cF8$–C	5.657(5)	5.657(5)	5.657(5)
	Y$_3$Ge$_5$	37.2	62.8	$oF64$–Y$_3$Ge$_5$	5.752(1)	17.288(5)	13.698(4)
	New Phase	40.2	59.8				
La	**La$_4$LiGe$_{10-x}$**	**29.3**	**70.7**	**$mS60$-y-La$_4$MgGe$_{10-x}$**	**8.844(2)**	**8.507(2)** β = 97.09(2)°	**17.827(3)**
	La$_2$LiGe$_6$	25.2	74.8	$mS36$–La$_2$AlGe$_6$	8.362(4)	8.864(5) β = 100.96(5)°	10.838(7)
Ce■	**Ce$_4$LiGe$_{10-x}$**	**30.5**	**69.5**	**$mS60$-y-La$_4$MgGe$_{10-x}$**	**8.701(8)**	**8.416(8)** β = 97.2(2)°	**17.70(3)**
	CeLiGe$_2$[a]	34.3	65.7	$oP16$–LiCaSi$_2$			
	CeGe$_{2-x}$	36.8	63.2	$oI12$–GdSi$_{2-x}$	4.242(6)	4.339(5)	14.08(1)
Pr■	**Pr$_4$LiGe$_{10-x}$**	**28.0**	**72.0**	**$mS60$-y-La$_4$MgGe$_{10-x}$**	**8.621(4)**	**8.326(4)** β = 97.17(9)°	**17.65(1)**
	PrLiGe$_2$[a]	34.6	65.4	$oP16$–LiCaSi$_2$			
Nd■	**Nd$_4$LiGe$_{10-x}$**	**29.1**	**70.9**	**$mS60$-y-La$_4$MgGe$_{10-x}$**	**8.566(3)**	**8.326(3)** β = 96.99(4)°	**17.65(4)**
	Nd$_2$LiGe$_6$#	25.3	74.7				
	Ge	–	100	$cF8$–C	5.652(1)		
	NdGe$_{2-x}$[a]	35.9	64.1	$oI12$–GdSi$_{2-x}$			
	NdLiGe2a	35.9	64.1	$oP16$–LiCaSi$_2$			

(continued)

Table 7.2 (continued)

R	Phases	Composition by (EDXS) [at.%]		Pearson symbol prototype	Lattice parameters [Å]		
		R	Ge		a	b	c
Sm	Sm_4LiGe_{10-x}	**28.6**	**71.4**	$mS60$-y-La_4MgGe_{10-x}	**8.490(1)**	**8.238(1)** $\beta = 96.98(1)°$	**17.46(2)**
	$SmLiGe_2$	34.6	65.4	$oP16$-$LiCaSi_2$	7.789(3)	3.902(1)	10.533(3)
Gd■	Gd_4LiGe_{10-x}	**30.0**	**70.0**	$mS60$-y-La_4MgGe_{10-x}	**8.41(1)**	**8.11(1)** $\beta = 97.1(3)°$	**17.36(3)**
	Gd_3Ge_5	38.9	61.1	$oF64$-Y_3Ge_5			
	$GdGe_{2-x}$#	34.8	65.2				
Tb	Tb_4LiGe_{10-x}	**29.3**	**70.7**	$mS60$-y-La_4MgGe_{10-x}	**8.284(4)**	**8.057(4)** $\beta = 96.91(8)°$	**17.29(2)**
	Ge	1.4	98.6	$cF8$-C	5.654(1)		
	$TbGe_{2-x}$#	32.4	67.6				
	Tb_3Ge_5[a]	37.1	62.9	$oF64$-Y_3Ge_5			
Dy■	Dy_4LiGe_{10-x}	**30.7**	**69.3**	$mS60$-y-La_4MgGe_{10-x}	**8.236(8)**	**7.989(8)** $\beta = 98.8(1)°$	**17.23(2)**
	$DyGe_{2-x}$#	34.6	65.4				
	$DyGe_2$#	32.7	67.3				
Ho■	$HoGe_{2-x}$	36.2	63.8	$oS28$-YGe_{2-x}	4.084(1)	29.719(4)	3.917(1)
	Ge	6.2	93.8	$cF8$-C	5.657(1)		

(continued)

Table 7.2 (continued)

R	Phases	Composition by (EDXS) [at.%]		Pearson symbol prototype	Lattice parameters [Å]		
		R	Ge		a	b	c
Er■	ErGe$_{2-x}$	36.1	63.9	$oS24$– DyGe$_{2-x}$	4.050(8)	29.50(3)	3.897(2)
	Ge	–	100	$cF8$–C	5.657(1)		
Tm	Tm$_2$LiGe$_4$	35.0	65.0	$mP8$– Li$_{0.5}$TmGe$_2$	4.005(2)	3.874(1) $\beta = \mathbf{104.21(8)}°$	8.124(5)
	Ge	–	100	$cF8$–C	5.652(2)		
Yb■	YbGe$_{2-x}$#	35.0	65.0				
	Yb$_3$Ge$_5$	38.9	61.1	$hP8$–Th$_3$Pd$_5$	6.843(3)	6.843(3)	4.176(3)
	Yb$_4$LiGe$_{10}$(?)#	31.4	68.6				
Lu	Ge	–	100	$cF8$–C	5.655(1)		
	LuGe$_{2-x}$	35.7	64.3	$oS12$–ZrSi$_2$	3.98(1)	15.655(6)	3.853(1)
	New Phase	18.4	81.6				

a Phases identified by XRPD analysis

Lattice parameters were not refined for these phases due to their low amount and/or strong peak overlapping

Phases identified only by EDXS analysis

■ Samples where TaGe$_2$ was detected

the Mg ones where no container contaminations were detected. The most common secondary phases are Ge and binary RGe_{2-x} crystallizing with different structures. Ternary phases, like R_2LiGe_6, $RLiGe_2$ and Tm_2LiGe_4, form only with Li whereas in the case of Mg no traces of other ternary compounds, except the wanted ones, were revealed.

Representative micrographs and indexed X–ray powder patterns for prepared samples, where the title compound formed/did not form, along the R series are reported in Figs. 7.1, 7.2, 7.3. SEM–BSE micrographs show many cracks and irregularities which are due to the high intermetallics brittleness.

Results clearly evidence the reduced reactivity of heavy rare earth metals on forming ternary germanides when combined with Mg and Li. Exceptions are the Eu–Mg–Ge (Figs. 7.1b and 7.2b) and Y–Mg/Li–Ge alloys. In the former, the absence of Eu_4MgGe_{10-x} is probably related to the well know particular chemistry of Eu, which is often divalent, with respect to all the other R; in the latter, Y behaves according to its

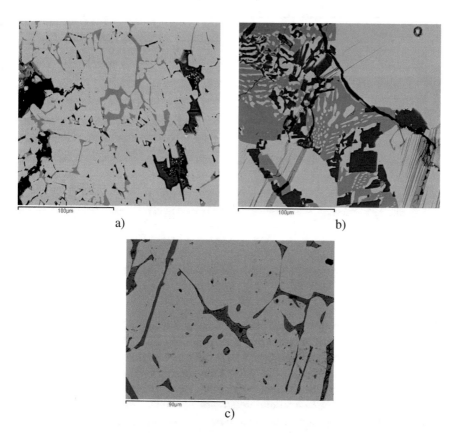

a)   b)

c)

Fig. 7.1 Micrographs of selected R–Mg–Ge (**a, b**) and R–Li–Ge (**c**) specimen: **a** Ce: bright phase Ce_4MgGe_{10-x}, grey phase Mg_2Ge, dark phase Ge; **b** Eu: bright phase $EuGe_2$, grey phase Mg_2Ge, dark phase Ge; **c** Tm: bright phase Tm_2LiGe_4, dark phase Ge

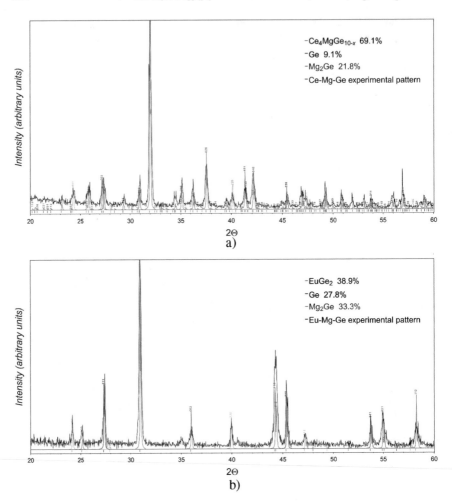

Fig. 7.2 X–ray powder patterns for Ce–Mg–Ge (**a**) and Eu–Mg–Ge (**b**). The 4:1:(10–x) stoichiometry phase was revealed only in (**a**)

dimensions which are close to that of heavy rare earth metals. The Tm–Li–Ge sample is also noteworthy. In fact, it is the alloy where the Tm_2LiGe_4 [2] (Figs. 7.1c and 7.3b) phase is reported to exist. Moreover, it is the only representative crystallizing with the monoclinic $mP8$– Tm_2LiGe_4 (SP: 11) [3]. Thus, further synthesis aiming to confirm its structure and testing the existence with other rare earth metals would be of great interest. Anyway, for the Li–containing samples, new syntheses have to be performed testing also other container materials aiming to avoid unwanted reactions between involved elements and crucible.

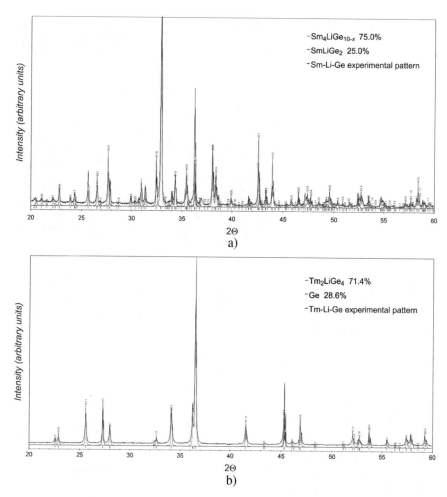

Fig. 7.3 X–ray powder patterns for Sm–Li–Ge (**a**) and Tm–Li–Ge (**b**). The 4:1:(10–x) stoichiometry phase was revealed only in (**a**)

7.3 Crystal Structure of Ce$_4$MgGe$_{10-x}$, Nd$_4$MgGe$_{10-x}$ and Nd$_4$LiGe$_{10-x}$

Single crystals of Ce$_4$MgGe$_{10-x}$, Nd$_4$MgGe$_{10-x}$ and Nd$_4$LiGe$_{10-x}$ compounds were selected from mechanically fragmented alloys with the aid of a light optical microscope operated in the dark field mode and analyzed by means of X–ray single crystal diffractometer (see paragraph 2.3.2.). Prior the selection, quality and composition of Nd$_4$LiGe$_{10-x}$ crystal were checked by means of SEM/EDXS. The polygonal crystal of good quality and size shown in Fig. 7.4 is the one subsequently selected for the X–ray single crystal experiments. Its EDXS composition of Nd$_{28.7}$Ge$_{71.3}$ is in very good agreement with that for the "Nd$_2$Ge$_5$" stoichiometry.

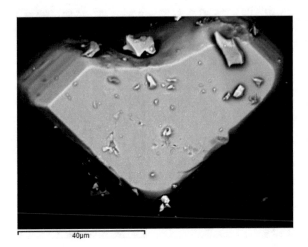

Fig. 7.4 SEM image of the Nd_4LiGe_{10-x} crystal selected from the mechanically fragmented sample for single crystal X–ray diffraction analysis

40μm

The crystal of Ce_4MgGe_{10-x} selected for X–ray analysis is an almost perfect non–merohedral twin composed of two domains, with one equal to about 1/3 of the other. Normally, twins of such type give problems on the preliminary stages of cell indexing (leading to unexpectedly high values of cell parameters) and on the space group determination (showing inconsistency with any known space group systematic absences) [4, 5]. In our case, the suspect of dealing with a twin came during refinement because the unit cell indexing was straightforward and possible space groups for the C–centered monoclinic cell (only the $h + k = 2n$ reflections were observed) were also correctly suggested: $C2$ (No. 5), Cm (No. 8) and $C2/m$ (No. 12). The lowest combined figure of merit is associated with the only $C2/m$ centrosymmetric space group, in which a preliminary structural model was obtained by direct methods. Reasoning on interatomic distances and thermal displacement parameters, 3 Ce, 8 Ge and 1 Mg positions were assigned for this model. But even with refined anisotropic thermal displacement parameters for all sites, the refinement sticks at R1/wR2 of 0.11/0.23. Moreover, the difference Fourier map shows additional intensive peaks in the vicinity of some Ge sites, that would have no physical sense in case of their full occupation. Associating these peaks to additional Ge positions and constraining occupation of corresponding pairs of Ge sites to be unity, residuals drop drastically to R1/wR2 of 0.039/0.084. The same disordered model was obtained testing the non–centrosymmetric $C2$ and Cm space groups. It is true that the disordering phenomenon often happens for Ge–rich compounds, but in this case such a model looks somewhat doubtful, being highly disordered. For this reason, a possible twinning was tested. The examination of collected data by XPREP [6] for a higher metric symmetry (with tolerance set to 0) did not reveal any possible pseudo–merohedral twinning to take place. After that, ROTAX [7], implemented in WinGX [8] package, and CELL_NOW [6], which are based on different approaches, gave the following 180° rotation twin law: $(-1\ 0\ 0;\ 0\ -1\ 0;\ 0.5\ 0\ 1)$. Subsequently two refinements were performed using hkl5 files generated by WinGX and SAINT&TWINABS [6] respectively. Both of them give very similar final discrepancy indices and flat differential Fourier maps for the same structural model. The final structural model was refined with anisotropic thermal displacement parameters for all atom sites (the noticeably higher U_{11} for Ge7 atom is associated with the fact that its site is partially occupied). The refined fractional contribution k of the second domain is 0.31.

The same procedure was then followed for the other crystals which turned out to be isostructural. For Nd₄LiGe₁₀₋ₓ, the Li species was put in the $4i$ position, analogously to Mg, leading to a residuals improvement. Selected crystallographic data and structure refinement parameters for the analysed crystals are listed in Tables 7.3, 7.4 and 7.5.

To illustrate the orientation of twin domains and the corresponding reciprocal plots of completely overlapped and non–overlapped hkl zones (Fig. 7.5), an alternative symmetry–related matrix, corresponding to a twofold rotation axis along the

Table 7.3 Crystallographic data for R_4MgGe_{10-x} (R = Ce and Nd) and Nd_4LiGe_{10-x} crystals together with experimental details about the structure solution. All compounds are isostructural

Empirical formula	Ce₄MgGe₁₀₋ₓ (x = 0.40)	Nd₄MgGe₁₀₋ₓ (x = 0.46)	Nd₄LiGe₁₀₋ₓ (x = 0.44)
EDXS composition	Ce₂₅.₃Mg₁₃.₇Ge₆₁.₀	Nd₂₄.₅Mg₁₂.₁Ge₆₃.₄	Nd₂₈.₇Ge₇₁.₃
Structure type	La₄MgGe₁₀₋ₓ		
Space group	C2/m (№ 12)		
Pearson symbol, Z	mS60–y, 4		
M_w, [g/mol]	1310.69	1327.17	1309.80
Unit cell dimensions:			
a, Å	8.7602(7)	8.5307(5)	8.611(1)
b, Å	8.5926(6)	8.2597(5)	8.337(1)
c, Å	17.585(1)	17.582(1)	17.622(2)
β, °	97.163(1)	96.971(1)	97.129(1)
V, Å³	1313.3(2)	1229.7(1)	1255.3(3)
Calc. density [g/cm³]	6.629	6.998	6.930
Twin law	[1 0 0; 0–1 0; –1/2 0–1]		
k (BASF)	0.31	0.52	0.47
Unique reflections	1441	2124	1354
Reflections $I > 2\sigma(I)$	1253 (R_{sigma} = 0.0095)	1176 (R_{sigma} = 0.0377)	1254
Data/parameters	1441/102	2124/92	1354/50
GOF on F² (S)	1.05	1.10	1.172
R indices [I > 2σ(I)]	R_1 = 0.0121; wR₂ = 0.0246	R_1 = 0.0301; wR₂ = 0.0642	R_1 = 0.0265; wR₂ = 0.0697
R indices [all data]	R_1 = 0.0166; wR₂ = 0.0295	R_1 = 0.0669; wR₂ = 0.0811	R_1 = 0.0295; wR₂ = 0.0716
$\Delta\rho_{fin}$ (max/min), [e/Å³]	0.82/–1.05	3.95/–3.34	3.68/–2.96

Table 7.4 Atomic coordinates and equivalent isotropic displacement parameters (Å^2) for R_4MgGe_{10-x} (R = Ce and Nd)

Atom	Wyck. site	Site	SOF $\neq$ 1	x/a	y/b	z/c	U_{iso}, Å^2
$Ce_4MgGe_{9.60}$							
Ce1	4i	m		0.59995(7)	0	0.39669(2)	0.0078(1)
Ce2	8j	1		0.04973(6)	0.25004(3)	0.19881(1)	0.0066(1)
Ce3	4i	m		0.09848(7)	0	0.39649(2)	0.0076(1)
Ge1	4i	m		0.06229(9)	0	0.06761(4)	0.0086(2)
Ge2	4i	m		0.47160(9)	0	0.06757(4)	0.0085(2)
Ge3	8j	1		0.26700(8)	0.20122(5)	0.06777(2)	0.0088(1)
Ge4	4i	m		0.1847(1)	0	0.73854(4)	0.0088(1)
Ge5	4i	m		0.3126(1)	0	0.24972(4)	0.0116(1)
Ge6	8j	1		0.3346(1)	0.24469(6)	0.33749(3)	0.0134(1)
Mg	4i	m		0.7762(4)	0	0.1038(1)	0.0131(4)
Ge7	8j	1	0.80	0.138(6)	0.2493(6)	0.5269(7)	0.046(6)
$Nd_4MgGe_{9.54}$							
Nd1	4i	m		0.5992(3)	0	0.39604(6)	0.0086(2)
Nd2	8j	1		0.0486(3)	0.2501(1)	0.19838(2)	0.0082(1)
Nd3	4i	m		0.0986(3)	0	0.39595(6)	0.0087(2)
Ge1	4i	m		0.038(1)	0	0.0685(2)	0.0539(2)
Ge2	4i	m		0.4829(6)	0	0.0683(2)	0.0315(1)
Ge3	8j	1		0.2672(8)	0.2155(4)	0.06905(8)	0.0506(8)
Ge4	4i	m		0.1876(7)	0	0.7460(1)	0.0109(4)
Ge5	4i	m		0.3115(6)	0	0.2502(1)	0.0108(4)
Ge6	8j	1		0.3359(6)	0.2491(2)	0.33946(6)	0.0129(2)
Mg	4i	m		0.775(3)	0	0.0952(8)	0.086(4)
Ge7	8j	1	0.77	0.158(2)	0.2498(7)	0.5229(7)	0.051(5)

monoclinic a axis (1 0 0 0 −1 0 −0.5 0 −1), was chosen. In fact, it equally well describes our twinning case. The effect of this matrix on the data confirms that only those with $h = 2n$ are affected by the twinning. Taking in consideration that only $h + k = 2n$ reflections are observed it results that a half of measured intensities are affected by twinning. The difference in intensities between overlapped/non–overlapped reflections are evident from the corresponding precession photos of $h2l$ and $h3l$ zones of reciprocal space shown in Fig. 7.5.

On the basis of interatomic distances analysis, the presence of a Ge covalent network can be guessed. In particular, it is composed by (3b)Ge corrugated layers, analogous to those described for R_2MGe_6 compounds, and by a complex framework where three zigzag chains are condensed together being composed by both (2b) and (3b)Ge. The latter is bridged to the former through the Mg/Li atoms with the R

Table 7.5 Atomic coordinates and equivalent isotropic displacement parameters (Å^2) for the Nd$_4$LiGe$_{10-x}$ crystal

Atom	Wyck. site	Site	SOF $\neq$ 1	x/a	y/b	z/c	U_{iso}, Å^2
Ce$_4$MgGe$_{9.60}$							
Nd1	4i	m		0.5987(2)	0	0.3963(1)	0.0063(4)
Nd2	8j	1		0.0488(2)	0.25018(1)	0.19775(2)	0.0048(1)
Nd3	4i	m		0.0973(2)	0	0.39612(9)	0.0053(4)
Ge1	4i	m		0.0525(3)	0	0.0683(2)	0.0074(6)
Ge2	4i	m		0.4828(3)	0	0.06809(2)	0.0091(6)
Ge3	8j	1		0.2672(3)	0.2061(1)	0.06817(6)	0.0085(2)
Ge4	4i	m		0.1869(5)	0	0.7420(2)	0.0086(5)
Ge5	4i	m		0.3114(5)	0	0.2492(2)	0.0080(5)
Ge6	8j	1		0.3356(4)	0.2469(2)	0.33972(5)	0.0104(2)
Li	4i	m		0.230(6)	0	0.888(2)	0.002(8)
Ge7	8j	1	0.80	0.0995(7)	0.2494(4)	0.5214(1)	0.103(3)

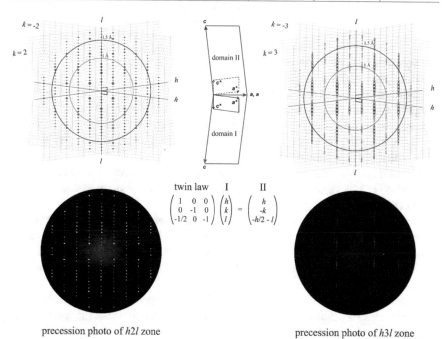

twin law   I   II

$$\begin{pmatrix} 1 & 0 & 0 \\ 0 & -1 & 0 \\ -1/2 & 0 & -1 \end{pmatrix} \begin{pmatrix} h \\ k \\ l \end{pmatrix} = \begin{pmatrix} h \\ -k \\ -h/2 - l \end{pmatrix}$$

precession photo of *h2l* zone   precession photo of *h3l* zone

Fig. 7.5 The upper part of the figure illustrates the reciprocal orientation of twin domains together with theoretical reciprocal plots of *h2l* (total overlapped) and *h3l* (non–overlapped) zones generated by XPREP. Domains I and II and corresponding hkl reflections are differentiated by red and green color respectively. Relations between direct/reciprocal lattice vectors lengths are not respected for clarity. The lower part shows experimental precession photos of *h2l* and *h3l* zones

Nd_4LiGe_{10-x}

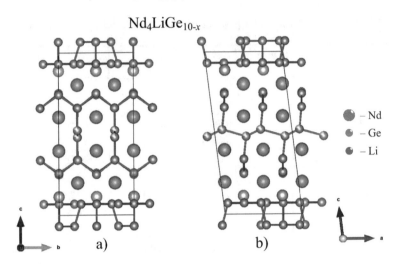

Fig. 7.6 Unit cell of Nd_4LiGe_{10-x}, chosen as a representative, shown in two different projections: along the a (**a**) and the b (**b**) axis. The red sticks represent Ge–Ge covalent bonds plotted on the basis of interatomic distances; analogies with R_2MGe_6 phases are clearly visible. The partially filled orange spheres corresponds to Ge7 positions whose site occupation factor is 0.80

species occupying the biggest cavities, again in a very similar arrangement as for the R_2MGe_6 germanides (Fig. 7.6).

These analogies suggest, on the basis of chemical bonding analysis presented in Chap. 4, that the title compounds could be reasonably described as germanolanthanates with $Li[R_4Ge_{10-x}]$ and $Mg[R_4Ge_{10-x}]$ general formula. Nevertheless, quantum chemistry investigation should be performed in order to definitively check the validity of this assumption.

The unit cell volumes were plotted as a function of the trivalent rare earth metal radii [9] (Fig. 7.7). Both lattice parameters obtained from single crystal and powder X–ray diffraction were considered.

In agreement with the lanthanide contraction, a linear decreasing trend is observed in both cases. The literature datum on La_4MgGe_{10-x} [1] fits very well in the general trend (Fig. 7.7a). The very good trends practically obtained on the basis of powder X–ray results, indicates the goodness of the indexing procedure and then the reliability of deduced unit cell parameters along he series.

To conclude, it is worth to note that although the proposed twinned model shows very good residuals and almost flat different Fourier maps, some data suggest that the obtained structure model should be improved. In particular, the shape and values for the Ge7 anisotropic displacement parameters and the presence of some additional very weak intensity reflections which have not been indexed. Some of them are visible to the left in Fig. 7.5. This kind of pattern suggest the presence of modulation. Further attempts, aiming to find a good twinned and modulated structure model for title compounds, will be performed.

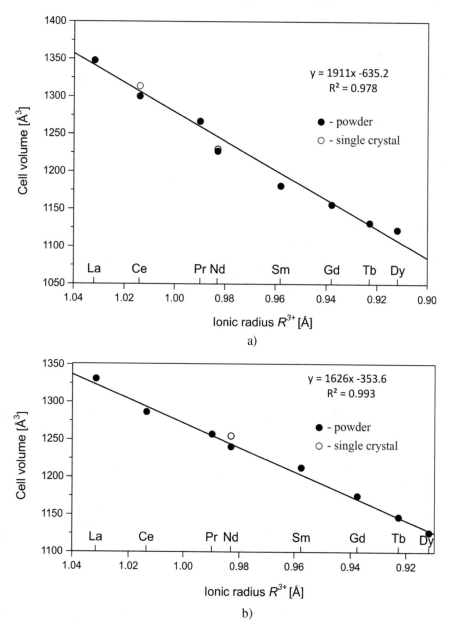

Fig. 7.7 Cell volumes (from both single crystal and powder data) of R_4MgGe$_{10-x}$ (**a**) and R_4LiGe$_{10-x}$ (**b**) compounds as a function of the R^{3+} ionic radius. The datum for La$_4$MgGe$_{10-x}$ was taken from [1]

7.4　Conclusions on the R_4MGe$_{10-x}$ (R = Rare Earth Metal; M = Li, Mg) Compounds

The existence and crystal structure of the R_4MGe$_{10-x}$ (R = rare earth metal; M = Li, Mg) ternary germanides was investigated along the whole R series. They turned out to form, with both Mg and Li, with R = La–Nd, Sm, Gd–Dy. Single crystals of Ce$_4$MgGe$_{10-x}$, Nd$_4$MgGe$_{10-x}$ and Nd$_4$LiGe$_{10-x}$ were selected for X–ray single crystal diffraction analysis. Results revealed that they are all non–merohedrally twins composed by two domains of similar dimensions, crystalizing with a monoclinic $mS60–y$ structure type, like the La$_4$MgGe$_{10-x}$ analogue. Interatomic distances analysis suggest the presence of (3b) and (2b)Ge atoms forming the same corrugated layer found within the R_2MGe$_6$ unit cells, and a complex framework composed by three zigzag chains connected through Ge–Ge bonds. Since also the coordination sphere of Mg/Li and R species are close to those revealed for R_2MGe$_6$, the description of this compounds as new families of germanolanthanates seems to be appropriate. Further study will be performed in order to check the chemical bonding and also to improve the structure model. In fact, low intensity peaks were detected suggesting the possibility to describe the R_4MGe$_{10-x}$ as modulated structures.

References

1. De Negri S, Solokha P, Skrobańska M, Proserpio DM, Saccone A (2014) J Solid State Chem 218:184–195. https://doi.org/10.1016/j.jssc.2014.06.036
2. Pavlyuk VV, Bodak OI, Sobolev AN (1991) Sov Phys Crystallogr 36:493–494
3. Villars P, Cenzual K (2018/19) Pearson's crystal data. ASM International, Ohio, USA
4. Clegg W (2009) Crystal structure analysis: principle and practice, 2nd edn. Oxford University Press, Oxford, UK
5. Müller P (ed) (2006) Crystal structure refinement: a crystallographer's guide to SHELXL. Oxford University Press, Oxford
6. Bruker (2014) APEX2, SAINT–Plus, XPREP, SADABS, CELL_NOW and TWINABS. Bruker AXS Inc., Madison, Wisconsin, USA
7. Cooper RI, Gould RO, Parsons S, Watkin DJ (2002) J Appl Crystallogr 35:168–174
8. Farrugia LJ (1999) J Appl Cryst 32:837–838

Chapter 8
Conclusions

Ternary R–M–Ge (R = rare earth metal; M = another metal) polar intermetallic germanides have been extensively studied on the basis of both experimental and theoretical methods.

Many synthetic efforts have been performed in order to check the existence and crystal structure of selected stoichiometry phases along the R series. In particular, the R_2PdGe_6 and R_2LiGe_6, the $R_2Pd_3Ge_5$, the R_4MgGe_{10-x} and R_4LiGe_{10-x} series have been investigated, employing also metal flux syntheses for some representatives. For the 2:1:6 stoichiometry phases with M = Pd, the wanted compounds formed with all the rare earths, but Sc and Eu. After several synthetic efforts, many of which driven by DTA measurement results, the La_2PdGe_6 turned out to be metastable. The use of Indium as metal flux was successful to stabilize it, allowing to select suitable single crystals for further X-ray single crystal analysis. The influence of In metal flux was tested also with Pr and Yb. The $oS72$-$Ce_2(Ga_{0.1}Ge_{0.9})_7$ structure type was assigned to all the R_2PdGe_6 synthesized by means of direct method and to the Yb-containing one isolated from In flux. On the other hand, the metal flux synthesized La_2PdGe_6 and Pr_2PdGe_6 show twinned crystals with $mS36$-La_2AlGe_6 structure, evidencing the tendency of In to stabilize, with early rare earths, the monoclinic isomorph. Therefore, the literature reported $oS18$-Ce_2CuGe_6 was never obtained. Invoking the vacancy ordering phenomenon, a compact Bärnighausen tree was constructed, with rigorous group-subgroup relations between the $oS20$-$SmNiGe_3$ aristotype and the three possible derivatives $oS72$, $mS36$ and $oS18$: their structural models are localized on separate branches of the tree, representing different symmetry reduction paths. The presence of a $t2$ reduction step on the path bringing to the $mS36$ model is coherent with the formation of twinned crystals. The experimental results were also confirmed by means of DFT total energy calculations. In order to propose a complete structure revision for all the R_2MGe_6 reported in the literature, the same DFT calculations were performed with R = La and M = Li, Cu, Ag, Au and Pt leading

© The Editor(s) (if applicable) and The Author(s), under exclusive license
to Springer Nature Switzerland AG 2020
R. Freccero, *Study of New Ternary Rare-Earth Intermetallic Germanides
with Polar Covalent Bonding*, Springer Theses,
https://doi.org/10.1007/978-3-030-58992-9_8

to the lowest energies for the $oS72$ and $mS36$ structure types. Experimental check of the first-principle results were tested for the R_2LiGe_6 (R = La-Nd, Sm) series, La_2CuGe_6 and La_2AgGe_6. In all cases, the $oS18$ type was never detected leading to the conclusion that this structure model has to be revised for all phases belonging to the R_2MGe_6 family which should adopt the monoclinic or the orthorhombic structure. In fact, La_2LiGe_6 forms twinned monoclinic $mS36$ crystals whereas the Cu- and Ag-containing analogues crystallized with the $oS72$ model.

The existence and crystal structure of $R_2Pd_3Ge_5$ phases was checked for R = La, Ce, Pr and two new representatives were detected and characterized: $Nd_2Pd_3Ge_5$ and $Yb_2Pd_3Ge_5$. All of them crystallize with the $oI40$-$U_2Co_3Si_5$ structure. This orthorhombic model, and its monoclinic $mS40$-$Lu_2Co_3Si_5$ derivative, were placed on a Bärnighausen tree where the 1:2:2 stoichiometry phase is the arystotype. In this tree, also the $LaPdGe_3$ is present as a hettotypes, laying on another branch. Structure similarities/differences were highlighted. Metal flux synthesis, involving Pd, Bi and In, were also performed. $La_2Pd_3Ge_5$ and $Nd_2Pd_3Ge_5$ crystals were obtained within Bi and Pb, respectively. For the latter, X-ray powder analysis performed on the sample after centrifugation at 500 °C, show the same structure as those obtained by direct method. With Indium, $Yb_2Pd_3Ge_5$ was not revealed. At its place, the new Yb_2PdGe_3, AlB_2 derivative, was isolated and characterized. It is worth to note that all the $R_2Pd_3Ge_5$ sample are almost single phase also when synthesized by arc/induction-melting, making them attractive for physical properties measurements. $Yb_2Pd_3Ge_5$ was the first one chosen for this purpose.

Electrical resistivity and Seebeck coefficient as a function of temperature indicate the $Yb_2Pd_3Ge_5$ metallic-like behavior, confirmed by the small zT figure of merit. Magnetization and susceptibility measurements as a function of magnetic field and temperature indicate that the studied compound has a paramagnetic behavior and suggest a nearly divalent Yb state. The Yb^{2+} dimensions are close to those of light R^{3+} cations suggesting that the size factor plays an important role for this series of compounds.

The last investigated series was the R_4MGe_{10-x} (M = Mg, Li) where the wanted phases were revealed with R = La-Nd, Sm, Gd–Dy. All the analyzed crystals are twinned with the $mS60$-La_4MgGe_{10-x} structure. Anyway, this model is probably an average structure and a more correct model would be obtained considering modulation.

$Lu_5Pd_4Ge_8$ and $Lu_3Pd_4Ge_4$ intermetallics were detected as secondary phases during the investigations on the Pd-containing alloys. Their crystal structure was solved ($mP34$–$Tm_5Pd_4Ge_8$ and $oI22$–$Gd_3Cu_4Ge_4$). The selected $Lu_5Pd_4Ge_8$ crystals turned out to be non-merohedral twins, definitively confirming the tendency of ternary polar germanides to form this kind of twinned crystals.

Careful crystal structures analysis shows the intriguing chemistry variety of the investigated intermetallics. All of them are represented in Fig. 8.1 on a Gibbs triangle together with the Ge–Ge covalent fragments deduced on the basis of interatomic distances. Moving from the Ge-richest phases to the center of the triangle, the number of Ge homocontacts decrease starting from $(2b)$ and $(3b)$Ge for the R_2MGe_6 and R_4MGe_{10-x}, to $(1b)Ge_2$ dumbbells and $(0b)$Ge for $Lu_3Pd_4Ge_4$.

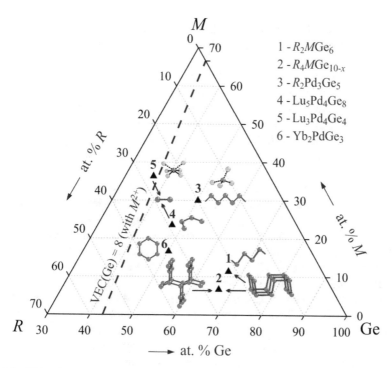

Fig. 8.1 Gibbs triangle (at.% Ge > 30%) reporting all the intermetallic phases investigated in this work. Ge–covalent fragments, deduced on the basis of interatomic distances, are shown for each of them. When (0b) Ge occur, the first surrounding Pd species are shown. As a guide to the eyes, the dashed line composed by all the alloys having a Valence Electron Concentration per Ge atom equal to 8 (assuming M to be divalent) is also represented

To a first approximation, this trend could be accounted for considering the charge transfer from metal atoms to Ge: a Ge decrease, which corresponds to a R/M increase, results in an higher number of transferred valence electrons. As a consequence, less Ge–Ge bonds have to be established to reach the octet. As an example, since the Zintl-Klemm approach is successful when applied to $R_2M^{(II)}Ge_6$ ($M^{(II)} = Mg$, Zn) to describe the Ge frameworks, the line composed by all the alloys having a Valence Electron Concentration per Ge atom equal to 8 is plotted in Fig. 8.1 (dashed line). The fact that $(1b)$Ge dumbbells occur within $Lu_3Pd_4Ge_4$ unit cell over the $VEC(Ge) = 8$, manifest the limits of the invoked approximation and the importance to study chemical bonding for some representatives on the basis of quantum chemical approaches.

A comparative chemical bonding analysis was performed for the La_2MGe_6 ($M =$ Li, Mg, Al, Zn, Cu, Ag, Pd) and Y_2PdGe_6 germanides by means of QTAIM, ELI-D, and their basin intersections. The presence of zigzag chains and corrugated layers of Ge atoms was confirmed. An approximate method (PSC0) was introduced, and applied for $M =$ Li, Mg, Al, Zn, to adapt the ELI-D valence electron count to general chemical expectations. It plays a decisive role to balance the Ge–La polar-covalent interactions against the Ge–M ones. All the compounds reveal significant deviations from the conceptual 8–N picture due to significant polar-covalent interactions with La and $M \neq$ Li, Mg atoms. For $M =$ Li, Mg title compounds were described as germanolanthanates, with $M[La_2Ge_6]$ formula. The relative Laplacian of ELI-D was discovered to reveal a chemically useful fine structure of the ELI-D distribution being related to polycentric bonding features. With the aid of this new tool, a consistent picture of Y–Pd and La–M interactions for all the title compounds was extracted. Deviations with respect to the 8–N rule were classified to be even larger than the ones found for CaGe, selected as a reference compound, especially when M is a transition metal.

DOS/COHP based chemical bonding studies were performed for $Lu_5Pd_4Ge_8$ and $Lu_3Pd_4Ge_4$ intermetallics, confirming the presence of covalently boned Ge atoms (Ge_2 dumbbells and cis-Ge_4 fragments). When compared with the corresponding Ge_2^{6-} and Ge_4^{10-} Zintl anions, they both show a reduced antibonding states population, caused by other bonding interactions with the surrounding Pd and Lu atoms. In fact, COHP curves and iCOHP values for Ge–Pd and Ge–Lu interactions confirm it, with the latter being of bonding type until E_F and above. This feature, also observed in other similar compounds (including the here studied La_2ZnGe_6 phase), confirm the relevance of Ge–R bonding in polar ternary germanides. The same is also true for the Lu–Pd interaction. Moreover, a short Pd–Pd distance (2.75Å) was found in $Lu_3Pd_4Ge_4$ and interpreted as a weak metal-metal bond. Preliminary ELI-D based results confirm the presence of Lu–Pd polar covalent bonds.

Even though accurate chemical bonding investigations were not performed for all the other compounds, a generalized scheme can be proposed. First of all, these compounds have to be treated as polar intermetallics where a partial charge transfer occurs. As a consequence, additional covalent interactions, even polycentric, take place beyond the Ge–Ge ones. The Ge–R heteropolar bonds seem to be of great importance in stabilizing polar ternary germanides being probably present in all considered cases. Only with $M =$ Li, Mg, the Ge–M interactions can be described as mainly ionic; as a consequence, the 2:1:6 and 4:1:(10-x) phases have been presented as germanolanthanates: $M[R_2Ge_6]$ and $M[R_4Ge_{10-x}]$. With the others M, the description of Ge–M interactions as heteropolar seems more appropriate. Finally, the importance of the M–R 2-center polar bonds is also clearly evidenced in this study playing a crucial role when M is a transition metal.

The synthesis, structural variety and fascinating chemical bonding scenarios for the studied intermetallics have been presented. These results constitute a step forward in the comprehension of composition-structure-properties relationships and a good playground for further studies on analogous systems.

Appendix

See Tables A3.1, A3.2, A3.3, A3.4, A3.5, A3.6, A3.7 and A3.8.
   See Tables A4.1, A4.2, A4.3, A4.4, A4.5, A4.6, A4.7, A4.8, A4.9, A4.10, A4.11, A4.12, A4.13 and A4.14.
   See Tables A5.1 and A5.2
   See Figs. A5.1 and A5.2
   See Figs. A6.1 and A6.2

Table A3.1 Spreadsheets generated to evaluate, on the basis of the patch areas, the exact number of La penultimate shell electrons to be given to each of the coordinating valence basins

	PEN_SHELL_TARGETED	ELI-D BASINS	BASINS NUMBER & NAME	PATCHES		Area_fraction	Elettrons to valence
1							
2	bsn-218-LP_La_53	Ge4-Ge6	bsn-111-Ge_12-Ge_82	4.133432		0.035360392	0.033963657
3	bsn-218-LP_La_53	Ge5-Ge6	bsn-131-Ge_123-Ge_126	4.19948		0.035925415	0.034506361
4	bsn-218-LP_La_53	Ge6-M	bsn-171-Ge_125-Mg_2	9.770695		0.083585652	0.080284018
5	bsn-218-LP_La_53	Ge6-M	bsn-170-Ge_82-Mg_15	9.849988		0.084263982	0.080935554
6	bsn-218-LP_La_53	Ge4-M	bsn-192-Ge_12-Mg_2	9.069657		0.077588461	0.074523717
7	bsn-218-LP_La_53	Ge5-M	bsn-188-Ge_13-Mg_8	9.173505		0.078476853	0.075377017
8	bsn-218-LP_La_53	Ge2-M/R-M	bsn-162-Ge_4-Mg_2	3.251368		0.027814573	0.026715898
9	bsn-218-LP_La_53	Ge2-M/R-M	bsn-160-Ge_42-Mg_8	3.761582		0.032179316	0.030908233
10	bsn-218-LP_La_53	LPGe3	bsn-203-LP_Ge_119	12.340577		0.105570297	0.10140027
11	bsn-218-LP_La_53	LPGe3	bsn-197-LP_Ge_47	13.187409		0.112814715	0.108358534
12	bsn-218-LP_La_53	LPGe3	bsn-211-LP_Ge_27	7.042098		0.060243243	0.057863635
13	bsn-218-LP_La_53	LPGe2	bsn-153-LP_Ge_4	9.868517		0.084422492	0.081087804
14	bsn-218-LP_La_53	LPGe2	bsn-141-LP_Ge_97	8.431782		0.072131613	0.069282414
15	bsn-218-LP_La_53	LPGe2	bsn-148-LP_Ge_108	8.672859		0.074193962	0.0712633
16	bsn-218-LP_La_53	Ge2-Ge2	bsn-65-Ge_97-Ge_108	1.422903		0.01217255	0.011691734
17	bsn-218-LP_La_53	Ge3-Ge3	bsn-81-Ge_47-Ge_27	0.745088		0.006374026	0.006122252
18	bsn-218-LP_La_53	Ge2-Ge3	bsn-85-Ge_4-Ge_27	0.994856		0.008510724	0.00817455
19	bsn-218-LP_La_53	Ge2-Ge3	bsn-92-Ge_108-Ge_119	0.978609		0.008371735	0.008041052
20	bsn-218-LP_La_53	R_core	bsn-15-La_53				
21					116.89441		0.9605
22	Electrons to valence:		0.9605 It must be positive			CHECK	CHECK
23						Below must be 1	Below must be 0
24						1	0.0000000

The case of La_2MgGe_6 was chosen as a representative

© The Editor(s) (if applicable) and The Author(s), under exclusive license
to Springer Nature Switzerland AG 2020
R. Freccero, *Study of New Ternary Rare-Earth Intermetallic Germanides with Polar Covalent Bonding*, Springer Theses,
https://doi.org/10.1007/978-3-030-58992-9

Table A3.2 Spreadsheets generated to evaluate, on the basis of the patch areas, the exact number of Mg penultimate shell electrons to be given to each of the coordinating valence basins

1	PEN_SHELL_TARGETED	ELI-D BASINS	BASINS NUMBER & NAME	PATCHES	Area_fraction	Electrons_to/from_valence
2	bsn-77-Mg_2	Ge4-M	bsn-192-Ge_12-Mg_2	8.167644	0.221126768	0.016297043
3	bsn-77-Mg_2	Ge5-M	bsn-184-Ge_53-Mg_10	8.194173	0.221845001	0.016349977
4	bsn-77-Mg_2	Ge6-M	bsn-169-Ge_19-Mg_2	7.17124	0.194150617	0.0143089
5	bsn-77-Mg_2	Ge6-M	bsn-171-Ge_125-Mg_2	7.171077	0.194146204	0.014308575
6	bsn-77-Mg_2	Ge2-M	bsn-162-Ge_4-Mg_2	6.232344	0.16873141	0.012435505
7						
8				36.93648		0.0737
9					CHECK	CHECK
10	Electrons to valence:		0.0737 It must be positive		Below must be 1	Below must be 0
11					1	0.0000000

The case of La$_2$MgGe$_6$ was chosen as a representative

Table A3.3 Spreadsheets generated to evaluate, on the basis of the patch areas, the exact number of valence electrons to be given, from each of the coordinating valence basins, to the (2b)Ge2 penultimate shell

1	PEN_SHELL_TARGETED	ELI-D BASINS	BASINS NUMBER & NAME	PATCHES	Area_fraction	Electrons_from_valence
2	bsn-55-Ge_4	Ge2-Ge3	bsn-85-Ge_4-Ge_27	3.164328	0.125503121	-0.033208126
3	bsn-55-Ge_4	Ge2-Ge2	bsn-68-Ge_63-Ge_44	3.559745	0.141186093	-0.03735784
4	bsn-55-Ge_4	LPGe2	bsn-143-LP_Ge_4	6.380744	0.253072148	-0.06696289
5	bsn-55-Ge_4	LPGe2	bsn-153-LP_Ge_4	6.380795	0.253074171	-0.066963426
6	bsn-55-Ge_4	Ge2-M	bsn-162-Ge_4-Mg_2	5.72753	0.227164468	-0.060107718
7				25.21314		-0.2646
8					CHECK	CHECK
9					Below must be 1	Below must be 0
10	Electrons from valence:		-0.2646 It must be negative		1	0.0000000

The case of La$_2$MgGe$_6$ was chosen as a representative

Table A3.4 Spreadsheets generated to evaluate, on the basis of the patch areas, the exact number of valence electrons to be given, from each of the coordinating valence basins, to the (2b)Ge3 penultimate shell

1	PEN_SHELL_TARGETED	ELI-D BASINS	BASINS NUMBER & NAME	PATCHES	Area_fraction	Electrons_from_valence
2	bsn-58-Ge_27	Ge2-Ge3	bsn-85-Ge_4-Ge_27	3.006423	0.119800905	-0.031483678
3	bsn-58-Ge_27	Ge3-Ge3	bsn-81-Ge_47-Ge_27	2.748624	0.109528048	-0.028783971
4	bsn-58-Ge_27	LPGe3	bsn-200-LP_Ge_27	9.575431	0.381564836	-0.100275239
5	bsn-58-Ge_27	LPGe3	bsn-211-LP_Ge_27	9.764683	0.389106211	-0.102257112
6						
7				25.09516		-0.2628
8					CHECK	CHECK
9					Below must be 1	Below must be 0
10	Electrons from valence:		-0.2628 It must be negative		1	0.0000000
11						

The case of La$_2$MgGe$_6$ was chosen as a representative

Table A3.5 Spreadsheets generated to evaluate, on the basis of the patch areas, the exact number of valence electrons to be given, from each of the coordinating valence basins, to the (3b)Ge4 penultimate shell

1	PEN_SHELL_TARGETED	ELI-D BASINS	BASINS NUMBER & NAME	PATCHES	Area_fraction	Electrons_from_valence
2	bsn-38-Ge_12	Ge4-M	bsn-192-Ge_12-Mg_2	8.801426	0.350795593	-0.092750355
3	bsn-38-Ge_12	Ge4-Ge6	bsn-111-Ge_12-Ge_82	5.602778	0.223308113	-0.059042665
4	bsn-38-Ge_12	Ge4-Ge6	bsn-137-Ge_12-Ge_55	5.700711	0.227211397	-0.060074693
5	bsn-38-Ge_12	Ge4-Ge4/5 (mS/oS)	bsn-93-Ge_12-Ge_16	4.984984	0.198684897	-0.052532287
6						
7				25.08990		-0.2644
8					CHECK	CHECK
9					Below must be 1	Below must be 0
10	Electrons from valence:		-0.26440 It must be negative		1	0.0000000

The case of La$_2$MgGe$_6$ was chosen as a representative

Table A3.6 Spreadsheets generated to evaluate, on the basis of the patch areas, the exact number of valence electrons to be given, from each of the coordinating valence basins, to the (3b)Ge5 penultimate shell

	PEN_SHELL_TARGETED	ELI-D BASINS	BASINS NUMBER & NAME	PATCHES	Area_fraction	Electrons_from_valence
1	PEN_SHELL_TARGETED	ELI-D BASINS	BASINS NUMBER & NAME	PATCHES	Area_fraction	Electrons_from_valence
2	bsn-42-Ge_53	Ge5-M	bsn-184-Ge_53-Mg_10	8.778557	0.349840454	-0.092147976
3	bsn-42-Ge_53	Ge5-Ge6	bsn-116-Ge_53-Ge_82	5.68863	0.226701598	-0.059713201
4	bsn-42-Ge_53	Ge5-Ge6	bsn-134-Ge_53-Ge_55	5.602754	0.223279293	-0.058811766
5	bsn-42-Ge_53	Ge5-Ge5/4 (mS/oS)	bsn-100-Ge_51-Ge_53	5.023089	0.200178655	-0.052727058
6						
7				25.09303		-0.2634
8					CHECK	CHECK
9					Below must be 1	Below must be 0
10	Electrons from valence:		-0.2634 It must be negative		1	0.0000000

The case of La_2MgGe_6 was chosen as a representative

Table A3.7 Spreadsheets generated to evaluate, on the basis of the patch areas, the exact number of valence electrons to be given, from each of the coordinating valence basins, to the (3b)Ge6 penultimate shell

	PEN_SHELL_TARGETED	ELI-D BASINS	BASINS NUMBER & NAME	PATCHES	Area_fraction	Electrons_from_valence
1	PEN_SHELL_TARGETED	ELI-D BASINS	BASINS NUMBER & NAME	PATCHES	Area_fraction	Electrons_from_valence
2	bsn-30-Ge_82	Ge6-M	bsn-170-Ge_82-Mg_15	9.074103	0.362252179	-0.096830007
3	bsn-30-Ge_82	Ge6-Ge4	bsn-111-Ge_12-Ge_82	5.395206	0.21538494	-0.057572394
4	bsn-30-Ge_82	Ge6-Ge5	bsn-116-Ge_53-Ge_82	5.382218	0.214866439	-0.057433799
5	bsn-30-Ge_82	Ge6-Ge6	bsn-107-Ge_102-Ge_82	5.197606	0.207496443	-0.055463799
6						
7				25.04913		-0.2673
8					CHECK	CHECK
9					Below must be 1	Below must be 0
10	Electrons from valence:		-0.26730 It must be negative		1	0.0000000

The case of La_2MgGe_6 was chosen as a representative

Table A3.8 Spreadsheets generated to correct the average electronic population of each valence basins on the basis of the values shown in Tables A3.1, A3.2, A3.3, A3.4, A3.5, A3.6 and A3.7

	Central-Atom	BASINS NUMBER	ELI-D BASINS	N(B)_before	R_correc	M_correc	Ge2_correc	Ge3_correc	Ge4_correc	Ge5_correc	Ge6_correc	N(B)_AFTER
2		85	Ge2-Ge3	1.14220	0.032431		-0.0332081	-0.03148				1.10994
3		68	Ge2-Ge2	1.33500	0.046767		-0.0747157					1.30705
4	Ge2_4	143	LPGe2	1.50760	0.221634		-0.0669629					1.66227
5		153	LPGe2	1.50780	0.221634		-0.0669634					1.66247
6		162	Ge2-M	1.60490	0.115248	0.01244	-0.0601077					1.67248
7		85	Ge2-Ge3	1.14220	0.032431		-0.03321	-0.0314837				1.10994
8		81	Ge3-Ge3	1.02360	0.024489			-0.0575679				0.99052
9	Ge3_27	200	LPGe3	2.20360	0.267622			-0.1002752				2.37095
10		211	LPGe3	2.23030	0.267622			-0.1022571				2.39567
11		192	Ge4-M	2.19410	0.149047	0.016297			-0.0927504			2.26669
12		111	Ge4-Ge6	1.92310	0.03396				-0.0590427		-0.05757	1.84045
13	Ge4_12	137	Ge4-Ge6	1.93270	0.033964				-0.0600747		-0.0575724	1.84902
14		93	Ge4-Ge4/5 (mS/oS)	1.80510					-0.0525323	-0.05273		1.69984
15		184	Ge5-M	2.19310	0.150754	0.01635				-0.092148		2.26806
16		116	Ge5-Ge6	1.92360	0.03451					-0.0597132	-0.05743	1.84096
17	Ge5_53	134	Ge5-Ge6	1.93120	0.034506					-0.0588118	-0.0574338	1.84946
18		100	Ge5-Ge5/4 (mS/oS)	1.80510						-0.05253	-0.0527271	1.69984
19		170	Ge6-M	2.22920	0.16122	0.0143089					-0.09683	2.30790
20		111	Ge6-Ge4	1.92310	0.03396					-0.05904	-0.0575724	1.84045
21	Ge6_82	116	Ge6-Ge5	1.92360	0.034506					-0.05971	-0.0574338	1.84096
22		107	Ge6-Ge6	1.85790							-0.1109276	1.74697

Table A4.1 Position-space bonding analysis for La_2MgGe_6 and La_2ZnGe_6 performed on the basis of FPLO calculations

Atoms (Ω)	Q^{form}	$Q^{eff.}(\Omega)$	ELIBON	$N_{val}^{ELI}(Ge)$	$N_{acc}^{ELI}(C^{Ge})$	$N_{cb}(Ge)$	$N_{lp}(Ge)$
Without PSCO							
La	+3	+1.3229	2.0086	–	–	–	–
(2b)Ge2	−2	−0.7421	−1.6912	5.1195	7.3062	2.1544	1.4826
(2b)Ge3	−2	−0.8456	−1.3322	5.0368	6.5838	1.4388	1.7990
(3b)Ge4	−1	− 0.2160	− 0.8656	4.4971	8.1584	3.5995	0.4488
(3b)Ge5	−1	−0.2996	−0.9371	4.4983	8.1387	3.6036	0.4474
(3b)Ge6	−1	−0.2606	−0.8182	4.5383	8.1204	3.4947	0.5218
Zn	+ 2	+ 0.0912	+ 2.5541	–	–	–	–
La	+3	+1.3428	+2.0317	–	–	–	–
(2b)Ge2	−2	−1.1289	−1.6276	5.3654	6.9663	1.5754	1.8950
(2b)Ge3	−2	−0.7474	−1.1900	5.0452	6.4121	1.3041	1.8706
(3b)Ge4	−1	−0.5229	−0.7349	4.8104	7.8450	3.0239	0.8933
(3b)Ge5	−1	−0.4870	−0.6981	4.8120	7.8483	3.0356	0.8882
(3b)Ge6	−1	−0.5265	−0.7853	4.8265	7.9396	3.0076	0.9095
Mg	+2	+1.4099	+1.9127	–	–	–	–
After PSCO							
La	+ 3	+ 1.3229	+ 3.0000	–	–	–	–
(2b)Ge2	−2	−0.7421	−2.2067	4.7395	7.3936	2.5639	1.0878
(2b)Ge3	−2	−0.8456	−1.9233	4.8355	6.9447	1.9680	1.4337
(3b)Ge4	−1	−0.2160	−0.9436	4.2123	7.8148	3.4225	0.3949
(3b)Ge5	−1	−0.2996	−0.9794	4.2944	7.8601	3.4630	0.4157
(3b)Ge6	−1	−0.2606	−0.9189	4.2561	7.8016	3.3486	0.4538
Zn	+ 2	+ 0.0912	+ 2.0000	–	–	–	–
La	+ 3	+ 1.3428	+ 3.0000	–	–	–	–
(2b)Ge2	−2	−1.1289	−2.2456	5.1184	7.3069	2.1232	1.4976
(2b)Ge3	−2	−0.7474	−1.7608	4.7390	6.6389	1.7903	1.4743
(3b)Ge4	−1	−0.5229	−0.9347	4.5167	7.5921	3.0186	0.7490
(3b)Ge5	−1	−0.4870	−0.9167	4.4805	7.5681	3.0040	0.7383
(3b)Ge6	−1	−0.5265	−0.9963	4.5189	7.6748	2.9911	0.7639
Mg	+ 2	+ 1.4099	+ 2.0000	–	–	–	–

Values before and after PSC0 correction, applied choosing atomic core charges corresponding to Ge^{4+}, Mg^{2+}, Zn^{2+} and La^{3+}, are given. Quantities referred to each species are reported

Table A4.2 Position-space bonding analysis for La$_2$MgGe$_6$ based on FPLO calculations

Central atom	ELI-Dbasin (B_i)	Atomicity	Before PSCO						After PSCO					
			$\bar{N}(B_i)$	$p(B_i^{Ge})$	cc	lpc	$p(B_i^{Mg})$	$\sum_j p(B_i^{La_j})$	$\bar{N}(B_i)$	$p(B_i^{Ge})$	cc	lpc	$p(B_i^{Mg})$	$\sum_j p(B_i^{La_j})$
(2b)Ge2	Ge2–Ge2	(2Ge, 4La)	1.2019	0.48939	1.00000	0.00000	–	0.02130	1.17620	0.47265	1.00000	0.00000	–	0.05462
	Ge2–Ge3	(2Ge, 4La)	0.9815	0.50912	0.98176	0.01824	–	0.01701	0.94632	0.49950	1.00000	0.00000	–	0.04378
	lp-Ge2	(Ge, 3La)	1.5966	0.89521	0.20957	0.79043	–	0.10328	1.75988	0.77454	0.45093	0.54907	–	0.22410
	lp-Ge2	(Ge, 3La)	1.5968	0.89517	0.20967	0.79033	–	0.10333	1.76008	0.77450	0.45099	0.54901	–	0.22413
	lp'-Ge2-Mg	(Ge, Mg, 4La)	1.5895	0.89261	0.21478	0.78522	0.05863	0.04869	1.66437	0.81923	0.36154	0.63846	0.06468	0.11603
(2b)Ge3	Ge3–Ge3	(2Ge, 4La)	0.8509	0.49336	1.00000	0.00000	–	0.01399	0.81071	0.48302	1.00000	0.00000	–	0.03471
	Ge3–Ge2	(2Ge, 4La)	0.9809	0.47375	1.00000	0.00000	–	0.01692	0.94570	0.45659	1.00000	0.00000	–	0.04370
	lp-Ge3	(Ge, 3La)	2.2841	0.90819	0.18362	0.81638	–	0.08686	2.43572	0.80155	0.39689	0.60311	–	0.19381
	lp-Ge3	(Ge, 3La)	2.2962	0.90859	0.18282	0.81718	–	0.08636	2.44681	0.80237	0.39526	0.60474	–	0.19289
(3b)Ge4	Ge4–Ge5	(2Ge)	1.8045	0.49704	1.00000	0.00000	0.00288(2)	–	1.68170	0.49894	1.00000	0.00000	0.00387(2)	–
	Ge4–Ge6	(2Ge, La)	1.9106	0.50099	0.99801	0.00199	0.00015	0.01141	1.81341	0.49182	1.00000	0.00000	0.00015	0.03062
	Ge4–Ge6	(2Ge, La)	1.9177	0.50284	0.99432	0.00568	0.00015	0.01136	1.81956	0.49354	1.00000	0.00000	0.00015	0.03052
	lp-Ge4	(Ge, Mg, 2La)	2.2122	0.90046	0.19908	0.80092	0.04850	0.04746	2.27740	0.82890	0.34220	0.65780	0.05550	0.11211
(3b)Ge5	Ge5–Ge4	(2Ge)	1.8045	0.49981	1.00000	0.00000	0.00288(2)	–	1.68170	0.49706	1.00000	0.00000	0.00387(2)	–
	Ge5–Ge6	(2Ge, La)	1.9132	0.50063	0.99875	0.00125	0.00020	0.01165	1.81340	0.49059	1.00000	0.00000	0.00020	0.03104
	Ge5–Ge6	(2Ge, La)	1.9197	0.50227	0.99547	0.00453	0.00020	0.01161	1.80780	0.48905	1.00000	0.00000	0.00020	0.03114
	lp-Ge5	(Ge, Mg, 2La)	2.2109	0.89923	0.20155	0.79845	0.04912	0.04893	2.26520	0.82591	0.34817	0.65183	0.05641	0.11504
(3b)Ge6	Ge6–Ge6	(2Ge)	1.8623	0.49944	1.00000	0.00000	0.00096(2)	–	1.73480	0.49931	1.00000	0.00000	0.00117(2)	–
	Ge6–Ge4	(2Ge, La)	1.9177	0.48553	1.00000	0.00000	0.00015	0.01136	1.81956	0.47567	1.00000	0.00000	0.00015	0.03052
	Ge6–Ge5	(2Ge, La)	1.9135	0.48748	1.00000	0.00000	0.00020	0.01165	1.80809	0.47956	1.00000	0.00000	0.00020	0.03113
	lp-Ge6	(Ge, Mg, 2La)	2.2461	0.90490	0.19020	0.80980	0.04118	0.05280	2.31244	0.83033	0.33934	0.66066	0.04721	0.12138

Characteristic quantities referred to each valence basin, before and after correction PSCO, are listed

Table A4.3 Position-space bonding analysis for La$_2$ZnGe$_6$ based on FPLO calculations

Central atom	ELI-D basin (B_i)	Atomicity	Before PSCO						After PSCO					
			$\bar{N}(B_i)$	$p(B_i^{Ge})$	cc	lpc	$p(B_i^{Zn})$	$\sum_j p(B_i^{La_j})$	$\bar{N}(B_i)$	$p(B_i^{Ge})$	cc	lpc	$p(B_i^{Zn})$	$\sum_j p(B_i^{La_j})$
(2b)Ge2	Ge2–Ge2	(2Ge;4La)	1.3711	0.48822	1.00000	0.00000	–	0.02370	1.31212	0.46791	1.00000	0.00000	–	0.06432
	Ge2–Ge3	(2Ge;4La)	1.0990	0.52211	0.95578	0.04422	–	0.02067	1.06161	0.49721	1.00000	0.00000	–	0.05037
	lp-Ge2	(Ge;Zn;3La)	2.4178	0.80164	0.39672	0.60328	0.11477	0.08218	2.50949	0.71681	0.56639	0.43361	0.08714	0.19470
	lp-Ge2	(Ge;Zn;3La)	2.4183	0.80143	0.39714	0.60286	0.11495	0.08216	2.51034	0.71659	0.56682	0.43318	0.08738	0.19464
(2b)Ge3	Ge3–Ge3	(2Ge;4La)	1.0016	0.49161	1.00000	0.00000	–	0.01637	0.98129	0.47952	1.00000	0.00000	–	0.04056
	Ge3–Ge2	(2Ge;4La)	1.0990	0.45842	1.00000	0.00000		0.02067	1.06161	0.45242	1.00000	0.00000	–	0.05037
	lp-Ge3	(Ge;3La)	2.2410	0.90129	0.19741	0.80259	–	0.09486	2.45039	0.79249	0.41503	0.58497	–	0.20400
	lp-Ge3	(Ge;3La)	2.2422	0.90126	0.19748	0.80252	–	0.09486	2.45145	0.79249	0.41501	0.58499	–	0.20396
(3b)Ge4	Ge4–Ge5	(2Ge)	2.1347	0.49829	1.00000	0.00000	–	–	2.01542	0.49332	1.00000	0.00000	–	–
	Ge4–Ge6	(2Ge;La)	1.9387	0.49249	1.00000	0.00000	–	0.01635	1.86180	0.47900	1.00000	0.00000	–	0.04266
	Ge4–Ge6	(2Ge;La)	1.9426	0.49346	1.00000	0.00000	–	0.01631	1.86527	0.47991	1.00000	0.00000	–	0.04258
	lp-Ge4	(Ge;Zn;2La)	2.1424	0.70948	0.58103	0.41897	0.24761	0.04210	2.07235	0.69055	0.61889	0.38111	0.19753	0.11110
(3b)Ge5	Ge5–Ge4	(2Ge)	2.1347	0.50166	0.99667	0.00333	–	–	2.01542	0.50663	0.98674	0.01326	–	–
	Ge5–Ge6	(2Ge;La)	1.9318	0.49560	1.00000	0.00000	–	0.01682	1.87435	0.48665	1.00000	0.00000	–	0.04311
	Ge5–Ge6	(2Ge;La)	1.9290	0.49487	1.00000	0.00000	–	0.01684	1.87166	0.48592	1.00000	0.00000	–	0.04317
	lp-Ge5	(Ge;Zn;2La)	2.1432	0.70707	0.58585	0.41415	0.24939	0.04311	2.09870	0.69172	0.61656	0.38344	0.19663	0.11122

(continued)

Table A4.3 (continued)

Central atom	ELI-D basin (B_i)	Atomicity	Before PSCO						After PSCO					
			$\bar{N}(B_i)$	$p(B_i^{Ge})$	cc	lpc	$p(B_i^{Zn})$	$\sum_j p(B_i^{La_j})$	$\bar{N}(B_i)$	$p(B_i^{Ge})$	cc	lpc	$p(B_i^{Zn})$	$\sum_j p(B_i^{La_j})$
(3b)Ge6	Ge6–Ge6	(2Ge)	2.1656	0.49991	1.00000	0.00000	–	–	2.02577	0.49990	1.00000	0.00000	–	–
	Ge6–Ge4	(2Ge;La)	1.9426	0.49006	1.00000	0.00000	–	0.01631	1.86527	0.47734	1.00000	0.00000	–	0.04258
	Ge6–Ge5	(2Ge;La)	1.9318	0.48747	1.00000	0.00000	–	0.01682	1.87435	0.47013	1.00000	0.00000	–	0.04311
	lp–Ge6	(Ge;Zn;2La)	2.0804	0.75082	0.49837	0.50163	0.20241	0.04604	2.03622	0.72284	0.55432	0.44568	0.15925	0.11717

Characteristic quantities referred to each valence basin, before and after correction PSCO, are listed

Table A4.4 Synthetic conditions applied in order to synthesize and isolate a La_2PdGe_6 single crystal

Nominal composition	Treatment	La_2PdGe_6	Comments
$La_{22.2}Pd_{11.1}Ge_{66.7}$	Induction or arc melting	–	$La(Pd,Ge)_{2-x}$ + $LaPdGe_3$ + Ge
$La_{22.2}Pd_{11.1}Ge_{66.7}$	Induction melting + annealing at 700 °C for 2 w	–	$La(Pd,Ge)_{2-x}$ + $LaPdGe_3$ + Ge
$La_{22.2}Pd_{11.1}Ge_{66.7}$	Cycle I	+	Thin border around $La(Pd,Ge)_{2-x}$
$La_{22.2}Pd_{11.1}Ge_{66.7}$	Cycle I + annealing at 500 °C for 2 w	+	Thin border around $La(Pd,Ge)_{2-x}$
$La_{22.2}Pd_{11.1}Ge_{66.7}$	Cycle I + annealing at 700 °C for 2 w	–	$La(Pd,Ge)_{2-x}$ + $LaPdGe_3$ + Ge
$La_{22.2}Pd_{11.1}Ge_{66.7}$	1–3 DTA cycles (max T = 1100 °C, heating/cooling rate = 5 °C/min) on arc melted sample	+	Border around $La(Pd,Ge)_{2-x}$ and around $LaPdGe_3$ (in some regions border is thick)
$La_{22.2}Pd_{11.1}Ge_{66.7}$	Annealing at 825 °C (30 min) during cooling in DTA of an arc melted sample	–	$La(Pd,Ge)_{2-x}$ + $LaPdGe_3$ + Ge
$La_{22.2}Pd_{11.1}Ge_{66.7}$	Annealing at 880 °C (30 min) during cooling in DTA of an arc melted sample	–	$La(Pd,Ge)_{2-x}$ + $LaPdGe_3$ + Ge
$La_{22.2}Pd_{11.1}Ge_{66.7}$	Arc melting + annealing at 1000 °C for 1 day	–	$La(Pd,Ge)_{2-x}$ + $LaPdGe_3$ + Ge (not clear microstructure)
$La_{22.2}Pd_{11.1}Ge_{66.7}$	Arc melting + annealing at 1000 °C for 1 day + annealing at 890 °C for 1 month	–	$La(Pd,Ge)_{2-x}$ + $LaPdGe_3$ + Ge
$La_{22.2}Pd_{11.1}Ge_{66.7}$	Arc melting + annealing at 1000 °C for 1 day + annealing at 890 °C for 1 month + annealing at 830 °C for 1 month	–	$La(Pd,Ge)_{2-x}$ + $LaPdGe_3$ + Ge
$La_{21}Pd_7Ge_{72}$	Synthesis in In flux cycle II (global composition measured in the region of sample after DTA with big yield of 2:1:6)	+	Crystals of $La(Pd,Ge)_{2-x}$ with border of 2:1:6
$La_{21}Pd_{15}Ge_{64}$	Synthesis in In flux cycle II (global composition chosen to avoid $La(Pd,Ge)_{2-x}$)	+	Small amount around $La(Pd,Ge)_{2-x}$ + In-Pd binary phases crystals

(continued)

Table A4.4 (continued)

Nominal composition	Treatment	La$_2$PdGe$_6$	Comments
La$_{21}$Pd$_7$Ge$_{72}$	Synthesis in In flux cycle II modified (without intermediate annealings)	+	Crystals of La(Pd,Ge)$_{2-x}$ with border of 2:1:6
La$_{21}$Pd$_7$Ge$_{72}$	Synthesis in In flux cycle III	+	Many small crystals of "pure" 2:1:6 (no border)

(+ means that the phase of interest has been detected in the sample, − means that it has not been detected)

Cycle I 25°C (10 °C/min) → 950 °C → 350 °C (-0.2 °C/min) → furnace switched off

Cycle II 25°C → (2 °C/min) → 1000 °C (5 h) → (-1.0 °C/min) → 850 °C(48 h) → (-0.3 °C/min) → 25 °C.

Cycle III 25°C → (10 °C/min) → 750 °C (24 h) → (-0.5 °C/min) → 25 °C

Table A4.5 Interatomic distances (<3.5 Å) for R_2PdGe_6

Atom 1	Atom 2	$R=$ Y d [Å]	$R=$ Ce d [Å]	$R=$ Pr d [Å]	$R=$ Nd d [Å]	$R=$ Er d [Å]	$R=$ Yb d [Å]	$R=$ Lu d [Å]
R	Ge3	2.957(1)	3.053(1)	3.054(1)	3.039(1)	2.925(1)	2.970(1)	2.901(1)
	Ge3	3.016(1)	3.092(1)	3.092(1)	3.075(1)	2.993(1)	3.034(1)	2.972(1)
	Ge2	3.057(1)	3.113(1)	3.113(1)	3.100(1)	3.036(1)	3.055(1)	3.023(1)
	Ge2	3.109(1)	3.144(1)	3.144(1)	3.137(1)	3.084(1)	3.105(1)	3.053(1)
	Ge6	3.144(1)	3.203(1)	3.203(1)	3.191(1)	3.121(1)	3.120(1)	3.107(1)
	Ge6	3.150(1)	3.205(1)	3.204(1)	3.194(1)	3.128(1)	3.131(1)	3.118(1)
	Pd	3.151(1)	3.218(1)	3.218(1)	3.206(1)	3.129(1)	3.149(1)	3.108(1)
	Pd	3.156(1)	3.219(1)	3.219(1)	3.207(1)	3.132(1)	3.150(1)	3.115(1)
	Ge5	3.153(1)	3.225(1)	3.225(1)	3.212(1)	3.124(1)	3.169(1)	3.103(1)
	Ge4	3.180(1)	3.232(1)	3.232(1)	3.221(1)	3.156(1)	3.178(1)	3.147(1)
	Ge2	3.124(1)	3.237(1)	3.237(1)	3.218(1)	3.098(1)	3.180(1)	3.095(1)
	Ge3	3.365(1)	3.368(1)	3.368(1)	3.365(1)	3.359(1)	3.351(1)	3.371(1)
Ge2	Ge2	2.448(1)	2.488(1)	2.488(2)	2.483(3)	2.446(1)	2.440(1)	2.439(2)
	Pd	2.448(1)	2.520(1)	2.520(1)	2.505(1)	2.430(1)	2.449(1)	2.418(1)
	Ge3	2.522(1)	2.532(1)	2.532(2)	2.528(3)	2.513(1)	2.490(1)	2.511(2)
	2R	3.057(1)	3.113(1)	3.113(1)	3.100(2)	3.036(1)	3.055(1)	3.023(1)
	2R	3.109(1)	3.144(1)	3.144(1)	3.137(2)	3.084(1)	3.105(1)	3.053(1)
	2R	3.124(1)	3.237(1)	3.237(1)	3.218(1)	3.098(1)	3.180(1)	3.095(1)
Ge3	Ge2	2.522(1)	2.532(1)	5.532(2)	2.528(3)	2.513(1)	2.490(1)	2.511(2)
	Ge3	2.653(1)	2.585(1)	2.585(2)	2.597(3)	2.668(1)	2.583(1)	2.697(2)
	2R	2.957(1)	3.053(1)	3.054(1)	3.039(2)	2.925(1)	2.970(1)	2.901(1)
	2R	3.016(1)	3.092(1)	3.092(1)	3.075(2)	2.993(1)	3.034(1)	2.972(1)
	2R	3.365(1)	3.368(1)	3.368(1)	3.365(1)	3.359(1)	3.351(1)	3.371(1)
Ge4	Ge5	2.503(1)	2.495(1)	2.495(1)	2.496(1)	2.505(1)	2.499(1)	2.506(1)
	Pd	2.493(1)	2.523(1)	2.523(1)	2.518(1)	2.481(1)	2.505(1)	2.476(2)
	2Ge6	2.547(1)	2.581(1)	2.581(1)	2.572(1)	2.533(1)	2.542(1)	2.525(1)
	2R	3.180(1)	3.232(1)	3.232(1)	3.221(1)	3.156(1)	3.159(1)	3.147(1)
	2Ge6	3.186(1)	3.241(1)	3.241(1)	3.232(1)	3.167(1)	3.178(1)	3.150(1)
Ge5	Ge4	2.503(1)	2.495(1)	2.495(1)	2.496(1)	2.505(1)	2.498(1)	2.506(1)
	Pd	2.500(1)	2.524(1)	2.524(1)	2.519(1)	2.488(1)	2.505(1)	2.482(1)
	2Ge6	2.547(1)	2.581(1)	2.581(1)	2.574(1)	2.531(1)	2.542(1)	2.524(1)
	2R	3.153(1)	3.225(1)	3.225(1)	3.212(1)	3.124(1)	3.159(1)	3.103(1)
	2Ge6	3.287(1)	3.241(1)	3.241(1)	3.230(1)	3.168(1)	3.169(1)	3.150(1)
Ge6	Ge6	2.493(1)	2.487(1)	2.487(1)	2.489(1)	2.496(1)	2.490(1)	2.495(1)
	Pd	2.514(1)	2.549(1)	2.549(1)	2.541(1)	2.499(1)	2.520(1)	2.491(1)

(continued)

Table A4.5 (continued)

Atom 1	Atom 2	R = Y d [Å]	R = Ce d [Å]	R = Pr d [Å]	R = Nd d [Å]	R = Er d [Å]	R = Yb d [Å]	R = Lu d [Å]
	Ge4	2.547(1)	2.581(1)	2.581(1)	2.572(1)	2.533(1)	2.542(1)	2.525(1)
	Ge5	2.547(1)	2.581(1)	2.581(1)	2.574(1)	2.531(1)	2.542(1)	2.525(1)
	R	3.144(1)	3.203(1)	3.203(1)	3.191(1)	3.121(1)	3.149(1)	3.107(1)
	R	3.150(1)	3.205(1)	3.204(1)	3.194(1)	3.128(1)	3.150(1)	3.118(1)
	Ge5	3.187(1)	3.241(1)	3.241(1)	3.230(1)	3.168(1)	3.159(1)	3.150(1)
	Ge4	3.186(1)	3.241(1)	3.241(1)	3.232(1)	3.167(1)	3.159(1)	3.150(1)
Pd	Ge2	2.448(1)	2.519(1)	2.520(1)	2.505(1)	2.430(1)	2.449(1)	2.418(1)
	Ge4	2.493(1)	2.523(1)	2.523(1)	2.518(1)	2.481(1)	2.498(1)	2.476(1)
	Ge5	2.500(1)	2.524(1)	2.524(1)	2.519(1)	2.488(1)	2.499(1)	2.482(1)
	2Ge6	2.514(1)	2.549(1)	2.549(1)	2.541(1)	2.499(1)	2.520(1)	2.491(1)
	2R	3.151(1)	3.218(1)	3.218(1)	3.206(1)	3.129(1)	3.120(1)	3.108(1)
	2R	3.156(1)	3.219(1)	3.219(1)	3.207(1)	3.132(1)	3.131(1)	3.115(1)

(R = Y, Ce, Pr, Nd, Er, Yb, Lu) crystallizing with the oS72 modification

Table A4.6 Interatomic distances (<3.5 Å) for R_2PdGe_6

Atom 1	Atom 2	R = La d [Å]	R = Pr d [Å]
R	Ge3	3.119(1)	3.064(1)
	Ge3	3.126(1)	3.089(1)
	Ge2	3.141(1)	3.112(1)
	Ge2	3.180(1)	3.153(1)
	Ge6	3.232(1)	3.205(1)
	Ge6	3.239(1)	3.206(1)
	Pd	3.259(1)	3.221(1)
	Pd	3.265(1)	3.223(1)
	Ge4	3.264(1)	3.230(1)
	Ge5	3.271(1)	3.237(1)
	Ge2	3.297(1)	3.243(1)
	Ge3	3.382(1)	3.373(1)
Ge2	Ge2	2.520(1)	2.512(1)
	Ge3	2.525(1)	2.511(1)
	Pd	2.570(1)	2.523(1)
	2R	3.141(1)	3.112(1)
	2R	3.180(1)	3.153(1)
	2R	3.297(1)	3.243(1)
Ge3	Ge2	2.525(1)	2.511(1)
	Ge3	2.582(1)	2.612(1)
	2R	3.119(1)	3.064(1)
	2R	3.126(1)	3.089(1)
	2R	3.382(1)	3.373(1)
Ge4	Ge4	2.488(1)	2.503(1)
	Pd	2.542(1)	2.526(1)
	2Ge6	2.598(1)	2.584(1)
	2R	3.265(1)	3.230(1)
	2Ge6	3.283(1)	3.245(1)
Ge5	Ge5	2.478(1)	2.494(1)
	Pd	2.542(1)	2.527(1)
	2Ge6	2.598(1)	2.583(1)
	2R	3.271(1)	3.237(1)
	2Ge6	3.282(1)	3.246(1)
Ge6	Ge6	2.476(1)	2.493(1)
	Pd	2.575(1)	2.553(1)
	Ge4	2.598(1)	2.584(1)
	Ge5	2.598(1)	2.583(1)

(continued)

Table A4.6 (continued)

Atom 1	Atom 2	$R =$ La d [Å]	$R =$ Pr d [Å]
	R	3.232(1)	3.205(1)
	R	3.239(1)	3.206(1)
	Ge5	3.282(1)	3.246(1)
	Ge4	3.283(1)	3.245(1)
Pd	Ge4	2.5420(2)	2.526(1)
	Ge5	2.5424(2)	2.527(1)
	Ge2	2.5697(3)	2.523(1)
	2Ge6	2.5754(2)	2.553(1)
	2R	3.2587(2)	3.221(1)
	2R	3.2647(2)	3.223(1)

($R =$ La, Pr) crystallizing with the mS36 modification

Table A4.7 Position-space bonding analysis for La$_2$LiGe$_6$

Central atom	ELI-D basin (B$_i$)	Atomicity	Before PSC0						After PSC0					
			$\bar{N}(B_i)$	$p(B_i^{Ge})$	cc	lpc	$p(B_i^{Li})$	$\sum_j p(B_i^{La_j})$	$\bar{N}(B_i)$	$p(B_i^{Ge})$	cc	lpc	$p(B_i^{Li})$	$\sum_j p(B_i^{La_j})$
(2b)Ge2	Ge2–Ge2	(2Ge, 4La)	1.45050	0.48459	1.00000	0.00000	–	0.03116	1.44598	0.46158	1.00000	0.00000	–	0.07720
	Ge2–Ge3	(2Ge, 4La)	1.33310	0.50784	0.98432	0.01568	–	0.02085	1.28425	0.49342	1.00000	0.00000	–	0.04995
	lp–Ge2	(Ge, 3La)	1.39640	0.88821	0.22357	0.77643	–	0.11150	1.53775	0.76509	0.46982	0.53018	–	0.23464
	lp–Ge2	(Ge, 3La)	1.39640	0.88821	0.22357	0.77643	–	0.11157	1.53775	0.76509	0.46982	0.53018	–	0.23471
	lp'–Ge2–Li	(Ge, Li, 4La)	1.47700	0.92349	0.15301	0.84699	0.01733	0.05877	1.54667	0.84262	0.31477	0.68523	0.02130	0.13570
(2b)Ge3	Ge3–Ge3	(2Ge, 4La)	1.13640	0.48803	1.00000	0.00000	–	0.02358	1.12058	0.47049	1.00000	0.00000	–	0.05885
	Ge3–Ge2	(2Ge, 4La)	1.33310	0.47138	1.00000	0.00000	–	0.02085	1.28425	0.45662	1.00000	0.00000	–	0.04995
	lp–Ge3	(Ge, 3La)	2.14160	0.90227	0.19546	0.80454	–	0.09357	2.30832	0.79449	0.41103	0.58897	–	0.20166
	lp–Ge3	(Ge, 3La)	2.14270	0.90232	0.19536	0.80464	–	0.09353	2.30944	0.79458	0.41084	0.58916	–	0.20156
(3b)Ge4	Ge4–Ge4	(2Ge)	1.97110	0.49992	1.00000	0.00000	–	–	1.84979	0.49992	1.00000	0.00000	–	–
	Ge4–Ge6	(2Ge, La)	1.79950	0.50414	0.99172	0.00828	–	0.01289	1.71756	0.49560	1.00000	0.00000	–	0.02972
	Ge4–Ge6	(2Ge, La)	1.80220	0.50488	0.99023	0.00977	–	0.01287	1.72008	0.49634	1.00000	0.00000	–	0.02968
	lp–Ge4	(Ge, Li, 2La)	2.01150	0.92220	0.15561	0.84439	0.01447	0.05996	2.08359	0.84843	0.30315	0.69685	0.01827	0.13004
(3b)Ge5	Ge5–Ge5	(2Ge)	1.93860	0.49969	1.00000	0.00000	–	–	1.83302	0.49967	1.00000	0.00000	–	–
	Ge5–Ge6	(2Ge, La)	1.79480	0.50290	0.99421	0.00579	–	0.01359	1.72055	0.49237	1.00000	0.00000	–	0.03448

(continued)

Table A4.7 (continued)

Central atom	ELI-D basin (B$_i$)	Atomicity	Before PSC0						After PSC0					
			$\overline{N}(B_i)$	$p(B_i^{Ge})$	cc	lpc	$p(B_i^{Li})$	$\sum_j p(B_i^{La_j})$	$\overline{N}(B_i)$	$p(B_i^{Ge})$	cc	lpc	$p(B_i^{Li})$	$\sum_j p(B_i^{La_j})$
	Ge5-Ge6	(2Ge, La)	1.79530	0.50298	0.99404	0.00596	–	0.01359	1.72075	0.49237	1.00000	0.00000	–	0.03447
	lp-Ge5	(Ge, Li, 2La)	2.04750	0.92552	0.14896	0.85104	0.01460	0.05714	2.11097	0.85184	0.29632	0.70368	0.01875	0.12676
(3b)Ge6	Ge6-Ge6	(2Ge)	1.98200	0.49995	1.00000	0.00000	–	–	1.86381	0.49994	1.00000	0.00000	–	–
	Ge6-Ge4	(2Ge, La)	1.79950	0.48297	1.00000	0.00000	–	0.01289	1.71756	0.47467	1.00000	0.00000	–	0.02972
	Ge6-Ge5	(2Ge, La)	1.79480	0.48345	1.00000	0.00000	–	0.01359	1.72055	0.47310	1.00000	0.00000	–	0.03448
	lp-Ge6	(Ge, Li, 2La)	2.13490	0.92506	0.14989	0.85011	0.01218	0.06141	2.20696	0.84982	0.30036	0.69964	0.01559	0.13327

Characteristic quantities referred to each valence basin, before and after correction PSC0, are listed

Table A4.8 Position-space bonding analysis for La_2MgGe_6

Central atom	ELI-D basin (B_i)	Atomicity	Before PSC0						After PSC0					
			$\bar{N}(B_i)$	$p(B_i^{Ge})$	cc	lpc	$p(B_i^{Mg})$	$\sum_j p(B_i^{La_j})$	$\bar{N}(B_i)$	$p(B_i^{Ge})$	cc	lpc	$p(B_i^{Mg})$	$\sum_j p(B_i^{La_j})$
(2b)Ge2	Ge2–Ge2	(2Ge, 4La)	1.3350	0.48787	1.00000	0.00000	–	0.02427	1.30705	0.46971	1.00000	0.00000	–	0.06057
	Ge2–Ge3	(2Ge, 4La)	1.1422	0.51698	0.96603	0.03397	–	0.02049	1.10994	0.50209	0.99582	0.00418	–	0.05030
	lp-Ge2	(Ge, 3La)	1.5076	0.88903	0.22194	0.77806	–	0.11031	1.66227	0.76602	0.46795	0.53205	–	0.23338
	lp-Ge2	(Ge, 3La)	1.5078	0.88898	0.22205	0.77795	–	0.11036	1.66247	0.76599	0.46801	0.53199	–	0.23341
	lp'-Ge2–Mg	(Ge, Mg, 4La)	1.6049	0.89688	0.20624	0.79376	0.05253	0.05047	1.67248	0.82470	0.35061	0.64939	0.05784	0.11734
(2b)Ge3	Ge3–Ge3	(2Ge, 4La)	1.0236	0.49101	1.00000	0.00000	–	0.01827	0.99052	0.47835	1.00000	0.00000	–	0.04360
	Ge3–Ge2	(2Ge, 4La)	1.1422	0.46244	1.00000	0.00000	–	0.02049	1.10994	0.44761	1.00000	0.00000	–	0.05030
	lp-Ge3	(Ge, 3La)	2.2036	0.90329	0.19341	0.80659	–	0.09194	2.37095	0.79724	0.40552	0.59448	–	0.19833
	lp-Ge3	(Ge, 3La)	2.2303	0.90441	0.19119	0.80881	–	0.09084	2.39567	0.79929	0.40142	0.59858	–	0.19628
(3b)Ge4	Ge4–Ge5	(2Ge)	1.8051	0.49731	1.00000	0.00000	–	–	1.69984	0.49721	1.00000	0.00000	–	–
	Ge4–Ge6	(2Ge, La)	1.9231	0.50029	0.99943	0.00057	–	0.01201	1.84045	0.49067	1.00000	0.00000	–	0.03101
	Ge4–Ge6	(2Ge, La)	1.9327	0.50277	0.99446	0.00554	–	0.01195	1.84902	0.49303	1.00000	0.00000	–	0.03086
	lp-Ge4	(Ge, Mg, 2La)	2.1941	0.90301	0.19397	0.80603	0.04384	0.04931	2.26669	0.83318	0.33365	0.66635	0.04963	0.11349
(3b)Ge5	Ge5–Ge4	(2Ge)	1.8051	0.49925	1.00000	0.00000	–	–	1.69984	0.49915	1.00000	0.00000	–	–
	Ge5–Ge6	(2Ge, La)	1.9236	0.50094	0.99813	0.00187	–	0.01227	1.84096	0.49099	1.00000	0.00000	–	0.03156

(continued)

Table A4.8 (continued)

Central atom	ELI-D basin (B_i)	Atomicity	Before PSC0						After PSC0					
			$\overline{N}(B_i)$	$p(B_i^{Ge})$	cc	lpc	$p(B_i^{Mg})$	$\sum_j p(B_i^{La_j})$	$\overline{N}(B_i)$	$p(B_i^{Ge})$	cc	lpc	$p(B_i^{Mg})$	$\sum_j p(B_i^{La_j})$
	Ge5–Ge6	(2Ge, La)	1.9312	0.50290	0.9942	0.0058	–	0.01222	1.84946	0.49333	1.00000	0.00000	–	0.03142
	lp–Ge5	(Ge, Mg, 2La)	2.1931	0.90146	0.19707	0.80293	0.04446	0.05098	2.26806	0.83104	0.33792	0.66208	0.05020	0.11576
(3b)Ge6	Ge6–Ge6	(2Ge)	1.8579	0.49954	1.00000	0.00000	–	–	1.74697	0.49952	1.00000	0.00000	–	–
	Ge6–Ge4	(2Ge, La)	1.9231	0.48749	1.00000	0.00000	–	0.01201	1.84045	0.47811	1.00000	0.00000	–	0.03101
	Ge6–Ge5	(2Ge, La)	1.9236	0.48654	1.00000	0.00000	–	0.01227	1.84096	0.47718	1.00000	0.00000	–	0.03156
	lp–Ge6	(Ge, Mg, 2La)	2.2292	0.90696	0.18608	0.81392	0.03741	0.05450	2.30790	0.83408	0.33184	0.66816	0.04234	0.12250

Characteristic quantities referred to each valence basin, before and after correction PSC0, are listed

Table A4.9 Position-space bonding analysis for La_2AlGe_6

Central atom	ELI-D basin (B_i)	Atomicity	Before PSC0						After PSC0					
			$\bar{N}(B_i)$	$p(B_i^{Ge})$	cc	lpc	$p(B_i^{Al})$	$\sum_j p(B_i^{La_j})$	$\bar{N}(B_i)$	$p(B_i^{Ge})$	cc	lpc	$p(B_i^{Al})$	$\sum_j p(B_i^{La_j})$
(2b)Ge2	Ge2–Ge2	(2Ge, 4La)	1.32370	0.48659	1.00000	0.00000	–	0.02681	1.31295	0.46532	1.0000	0.00000	–	0.06944
	Ge2–Ge3	(2Ge, 4La)	1.05410	0.53894	0.92211	0.07789	–	0.01755	1.00905	0.52779	0.94441	0.05559	–	0.04217
	lp–Ge2	(Ge, 3La)	1.49580	0.88989	0.22022	0.77978	–	0.10770	1.64591	0.76731	0.46537	0.53463	–	0.23050
	lp–Ge2	(Ge, 3La)	1.49600	0.88984	0.22032	0.77968	–	0.10769	1.64611	0.76728	0.46544	0.53456	–	0.23047
	lp'–Ge2–Al	(Ge, Al, 4La)	1.95660	0.79234	0.41531	0.58469	0.15624	0.05131	2.06073	0.72253	0.55494	0.44506	0.15582	0.12156
(2b)Ge3	Ge3–Ge3	(2Ge, 4La)	1.11810	0.48788	1.00000	0.00000	–	0.02468	1.10214	0.47069	1.00000	0.00000	–	0.05907
	Ge3–Ge2	(2Ge, 4La)	1.05410	0.44360	1.00000	0.00000	–	0.01755	1.00905	0.43014	1.00000	0.00000	–	0.04217
	lp–Ge3	(Ge, 3La)	2.23400	0.89736	0.20528	0.79472	–	0.09861	2.42044	0.78604	0.42792	0.57208	–	0.21024
	lp–Ge3	(Ge, 3La)	2.23920	0.89769	0.20463	0.79537	–	0.09607	2.42503	0.78653	0.42695	0.57305	–	0.20985
(3b)Ge4	Ge4–Ge4	(2Ge)	1.94750	0.49997	1.00000	0.00000	–	–	1.82948	0.50000	1.00000	0.00000	–	–
	Ge4–Ge6	(2Ge, La)	1.98540	0.50463	0.99073	0.00927	–	0.01662	1.91182	0.49232	1.00000	0.00000	–	0.04008
	Ge4–Ge6	(2Ge, La)	1.98970	0.50565	0.98869	0.01131	–	0.01659	1.91566	0.49328	1.00000	0.00000	–	0.04000
	lp–Ge4	(Ge, Al, 2La)	2.24960	0.80085	0.39829	0.60171	0.15736	0.04018	2.32162	0.74259	0.51481	0.48519	0.15930	0.09656
(3b)Ge5	Ge5–Ge5	(2Ge)	1.96840	0.50000	1.00000	0.00000	–	–	1.86013	0.50000	1.00000	0.00000	–	–
	Ge5–Ge6	(2Ge, La)	1.97130	0.50084	0.99833	0.00167	–	0.01770	1.90877	0.48656	1.00000	0.00000	–	0.04647
	Ge5–Ge6	(2Ge, La)	1.97410	0.50160	0.99681	0.00319	–	0.01768	1.91151	0.48735	1.00000	0.00000	–	0.04641
	lp–Ge5	(Ge, Al, 2La)	2.26920	0.80077	0.39847	0.60153	0.15693	0.04045	2.33117	0.74273	0.51455	0.48545	0.16035	0.09513

(continued)

Table A4.9 (continued)

| Central atom | ELI-D basin (B_i) | Atomicity | Before PSC0 | | | | | | After PSC0 | | | | | |
			$\bar{N}(B_i)$	$p(B_i^{Ge})$	cc	lpc	$p(B_i^{Al})$	$\sum_j p(B_i^{La_j})$	$\bar{N}(B_i)$	$p(B_i^{Ge})$	cc	lpc	$p(B_i^{Al})$	$\sum_j p(B_i^{La_j})$
(3b)Ge6	Ge6–Ge6	(2Ge)	2.04410	0.49993	1.00000	0.00000	–	–	1.92464	0.49992	1.00000	0.00000	–	–
	Ge6–Ge4	(2Ge, La)	1.98970	0.47766	1.00000	0.00000	–	0.01659	1.91566	0.46661	1.00000	0.00000	–	0.04000
	Ge6–Ge5	(2Ge, La)	1.97410	0.48073	1.00000	0.00000	–	0.01768	1.91151	0.46624	1.00000	0.00000	–	0.04641
	lp–Ge6	(Ge, Al, 2La)	2.18390	0.82760	0.34480	0.65520	0.12427	0.04735	2.25757	0.76287	0.47426	0.52574	0.12591	0.11047

Characteristic quantities referred to each valence basin, before and after correction PSC0, are listed

Table A4.10 Position-space bonding analysis for La_2ZnGe_6

Central atom	ELI-D basin (B_i)	Atomicity	Before PSC0						After PSC0					
			$\bar{N}(B_i)$	$p(B_i^{Ge})$	cc	lpc	$p(B_i^{Zn})$	$\sum_j p(B_i^{La_j})$	$\bar{N}(B_i)$	$p(B_i^{Ge})$	cc	lpc	$p(B_i^{Zn})$	$\sum_j p(B_i^{La_j})$
(2b)Ge2	Ge2–Ge2	(2Ge;4La)	1.5009	0.48631	1.00000	0.00000	–	0.02705	1.47698	0.46554	1.00000	0.00000	–	0.06872
	Ge2–Ge3	(2Ge;4La)	1.2643	0.53112	0.93775	0.06225	–	0.02341	1.23426	0.51368	0.97263	0.02737	–	0.05811
	lp-Ge2	(Ge;Zn;3La)	2.3347	0.79603	0.40793	0.59207	0.11603	0.08721	2.47000	0.71474	0.57053	0.42947	0.08805	0.19652
	lp-Ge2	(Ge;Zn;3La)	2.3350	0.79610	0.40779	0.59221	0.11597	0.08719	2.47075	0.71459	0.57081	0.42919	0.08825	0.19646
(2b)Ge3	Ge3–Ge3	(2Ge;4La)	1.1586	0.48964	1.00000	0.00000	–	0.02037	1.12484	0.47513	1.00000	0.00000	–	0.04938
	Ge3–Ge2	(2Ge;4La)	1.2643	0.44546	1.00000	0.00000	–	0.02341	1.23426	0.42821	1.00000	0.00000	–	0.05811
	lp-Ge3	(Ge;3La)	2.1681	0.89622	0.20756	0.79244	–	0.10009	2.35070	0.78448	0.43105	0.56895	–	0.21212
	lp-Ge3	(Ge;3La)	2.1812	0.89685	0.20631	0.79369	–	0.09953	2.36280	0.78558	0.42884	0.57116	–	0.21108
(3b)Ge4	Ge4–Ge5	(2Ge)	2.0749	0.49935	1.00000	0.00000	–	–	1.95398	0.49937	1.00000	0.00000	–	–
	Ge4–Ge6	(2Ge;La)	1.9335	0.49532	1.00000	0.00000	–	0.01681	1.86747	0.48239	1.00000	0.00000	–	0.04243
	Ge4–Ge6	(2Ge;La)	1.9340	0.49545	1.00000	0.00000	–	0.01680	1.86791	0.48251	1.00000	0.00000	–	0.04242
	lp-Ge4	(Ge;Zn;2La)	2.1642	0.71745	0.56510	0.43490	0.23792	0.04371	2.11533	0.69551	0.60898	0.39102	0.19264	0.11091
(3b)Ge5	Ge5–Ge4	(2Ge)	2.0749	0.50065	0.99884	0.00116	–	–	1.95398	0.50063	0.99888	0.00112	–	–
	Ge5–Ge6	(2Ge;La)	1.9347	0.49610	1.00000	0.00000	–	0.01721	1.86901	0.48295	1.00000	0.00000	–	0.04315
	Ge5–Ge6	(2Ge;La)	1.9358	0.49638	1.00000	0.00000	–	0.01720	1.87015	0.48326	1.00000	0.00000	–	0.04312
	lp-Ge5	(Ge;Zn;2La)	2.1658	0.71313	0.57374	0.42626	0.24139	0.04497	2.11666	0.69126	0.61747	0.38253	0.19582	0.11239

(continued)

Table A4.10 (continued)

Central atom	ELI-D basin (B_i)	Atomicity	Before PSC0						After PSC0					
			$\bar{N}(B_i)$	$p(B_i^{Ge})$	cc	lpc	$p(B_i^{Zn})$	$\sum_j p(B_i^{La_j})$	$\bar{N}(B_i)$	$p(B_i^{Ge})$	cc	lpc	$p(B_i^{Zn})$	$\sum_j p(B_i^{La_j})$
(3b)Ge6	Ge6–Ge6	(2Ge)	2.1124	0.49995	1.00000	0.00000	–	–	1.98777	0.49995	1.00000	0.00000	–	–
	Ge6–Ge4	(2Ge;La)	1.9335	0.48787	1.00000	0.00000	–	0.01681	1.86747	0.47519	1.00000	0.00000	–	0.04243
	Ge6–Ge5	(2Ge;La)	1.9358	0.48636	1.00000	0.00000	–	0.01720	1.87015	0.47357	1.00000	0.00000	–	0.04312
	lp–Ge6	(Ge;Zn;2La)	2.1052	0.75679	0.48641	0.51359	0.19438	0.04874	2.07987	0.72566	0.54868	0.45132	0.15515	0.11910

Characteristic quantities referred to each valence basin, before and after correction PSC0, are listed

Table A4.11 Position-space bonding analysis for La$_2$CuGe$_6$

Central atom	ELI-D basin (B$_i$)	Atomicity	Before PSC0				After PSC0 (Cu$^+$)				After PSC0 (Cu^{2+})			
			$\bar{N}(B_i)$	$p(B_i^{Ge})$	cc	lpc	$\bar{N}(B_i)$	$p(B_i^{Ge})$	cc	lpc	$\bar{N}(B_i)$	$p(B_i^{Ge})$	cc	lpc
(2b)Ge2	Ge2–Ge2	(2Ge;4La)	1.5785	0.48470	1.00000	0.00000	1.56350	0.46178	1.00000	0.00000	1.56350	0.46178	1.00000	0.00000
	Ge2–Ge3	(2Ge;4La)	1.4765	0.52692	0.94616	0.05384	1.44920	0.50710	0.98580	0.01420	1.44920	0.50710	0.98580	0.01420
	lp–Ge2	(Ge;Cu;3La)	2.2113	0.79121	0.41758	0.58242	2.27186	0.73129	0.53742	0.46258	2.37943	0.69823	0.60354	0.39646
	lp–Ge2	(Ge;Cu;3La)	2.2114	0.79117	0.41765	0.58235	2.27208	0.73137	0.53727	0.46273	2.37984	0.69825	0.60350	0.39650
(2b)Ge3	Ge3–Ge3	(2Ge;4La)	1.2608	0.49032	1.00000	0.00000	1.21697	0.47303	1.00000	0.00000	1.21697	0.47303	1.00000	0.00000
	Ge3–Ge2	(2Ge;4La)	1.4765	0.44348	1.00000	0.00000	1.44920	0.41925	1.00000	0.00000	1.44920	0.41925	1.00000	0.00000
	lp–Ge3	(Ge;3La)	2.1116	0.89458	0.21084	0.78916	2.28179	0.77724	0.44552	0.55448	2.28179	0.77724	0.44552	0.55448
	lp–Ge3	(Ge;3La)	2.1169	0.89475	0.21050	0.78950	2.28676	0.77763	0.44473	0.55527	2.28676	0.77763	0.44473	0.55527
(3b)Ge4	Ge4–Ge5	(2Ge)	2.1183	0.49993	1.00000	0.00000	1.98834	0.50260	0.99481	0.00519	1.98834	0.50260	0.99481	0.00519
	Ge4–Ge6	(2Ge;La)	1.9130	0.48813	1.00000	0.00000	1.85252	0.47534	1.00000	0.00000	1.85252	0.47534	1.00000	0.00000
	Ge4–Ge6	(2Ge;La)	1.9157	0.48891	1.00000	0.00000	1.85503	0.47611	1.00000	0.00000	1.85503	0.47611	1.00000	0.00000
	lp–Ge4	(Ge;Cu;2La)	2.0992	0.72308	0.55383	0.44617	1.92108	0.74895	0.50209	0.49791	2.12962	0.67561	0.64877	0.35123
(3b)Ge5	Ge5–Ge4	(2Ge)	2.1183	0.50007	0.99981	0.00019	1.98834	0.49740	1.00000	0.00000	1.98834	0.49740	1.00000	0.00000
	Ge5–Ge6	(2Ge;La)	1.9104	0.48922	1.00000	0.00000	1.84123	0.47350	1.00000	0.00000	1.84123	0.47350	1.00000	0.00000
	Ge5–Ge6	(2Ge;La)	1.9125	0.48978	1.00000	0.00000	1.84314	0.47404	1.00000	0.00000	1.84314	0.47404	1.00000	0.00000
	lp–Ge5	(Ge;Cu;2La)	2.1083	0.71982	0.56036	0.43964	1.91028	0.74557	0.50887	0.49113	2.12386	0.67059	0.65882	0.34118

(continued)

Table A4.11 (continued)

Central atom	ELI-D basin (B_i)	Atomicity	Before PSC0				After PSC0 (Cu^+)				After PSC0 (Cu^{2+})			
			$\bar{N}(B_i)$	$p(B_i^{Ge})$	cc	lpc	$\bar{N}(B_i)$	$p(B_i^{Ge})$	cc	lpc	$\bar{N}(B_i)$	$p(B_i^{Ge})$	cc	lpc
(3b)Ge6	Ge6–Ge6	(2Ge)	2.1443	0.49988	1.00000	0.00000	2.00774	0.49988	1.00000	0.00000	2.00774	0.49988	1.00000	0.00000
	Ge6–Ge4	(2Ge;La)	1.9130	0.49289	1.00000	0.00000	1.85252	0.47655	1.00000	0.00000	1.85252	0.47655	1.00000	0.00000
	Ge6–Ge5	(2Ge;La)	1.9104	0.49157	1.00000	0.00000	1.84123	0.47766	1.00000	0.00000	1.84123	0.47766	1.00000	0.00000
	lp–Ge6	(Ge;Cu;2La)	2.0378	0.74708	0.50584	0.49416	1.87813	0.76329	0.47342	0.52658	2.05940	0.69611	0.60779	0.39221

Characteristic quantity referred to each valence basin, before and after correction PSC0, are listed

Table A4.12 Position-space bonding analysis for La_2AgGe_6

Central atom	ELI-D basin (B_i)	Atomicity	(B_i)	$p(B_i^{Ge})$	cc	lpc
(2b)Ge2	Ge2–Ge2	(2Ge;4La)	1.5701	0.48526	1.00000	0.00000
	Ge2–Ge3	(2Ge;4La)	1.3721	0.52620	0.94760	0.05240
	lp–Ge2	(Ge;Ag;3La)	2.1605	0.81814	0.36371	0.63629
	lp–Ge2	(Ge;Ag;3La)	2.16050	0.81819	0.36362	0.63638
(2b)Ge3	Ge3–Ge3	(2Ge;4La)	1.1803	0.48954	1.00000	0.00000
	Ge3–Ge2	(2Ge;4La)	1.3721	0.44734	1.00000	0.00000
	lp–Ge3	(Ge;3La)	2.1500	0.89656	0.20688	0.79312
	lp–Ge3	(Ge;3La)	2.1589	0.89698	0.20603	0.79397
(3b)Ge4	Ge4–Ge5	(2Ge)	2.1350	0.49789	1.00000	0.00000
	Ge4–Ge6	(2Ge;La)	1.9509	0.49526	1.00000	0.00000
	Ge4–Ge6	(2Ge;La)	1.9533	0.49588	1.00000	0.00000
	lp–Ge4	(Ge;Ag;2La)	1.9052	0.74165	0.51669	0.48331
(3b)Ge5	Ge5–Ge4	(2Ge)	2.1350	0.50211	0.99578	0.00422
	Ge5–Ge6	(2Ge;La)	1.9462	0.49486	1.00000	0.00000
	Ge5–Ge6	(2Ge;La)	1.9470	0.49507	1.00000	0.00000
	lp–Ge5	(Ge;Ag;2La)	1.9235	0.74229	0.51541	0.48459
(3b)Ge6	Ge6–Ge6	(2Ge)	2.1336	0.49977	1.00000	0.00000
	Ge6–Ge4	(2Ge;La)	1.9533	0.48861	1.00000	0.00000
	Ge6–Ge5	(2Ge;La)	1.9462	0.48926	1.00000	0.00000
	lp–Ge6	(Ge;Ag;2La)	1.9201	0.77454	0.45091	0.54909

Characteristic quantity referred to each valence basin are listed

Table A4.13 Position–space bonding analysis for La$_2$PdGe$_6$

Central Atom	ELI-D basin (B_i)	Atomicity	(B_i)	$p(B_i^{Ge})$	cc	lpc
(2b)Ge2	Ge2–Ge2	(2Ge;4La)	1.5879	0.48290	1.00000	0.00000
	Ge2–Ge3	(2Ge;4La)	1.5622	0.53175	0.93650	0.06350
	lp–Ge2	(Ge;Pd;3La)	2.0539	0.80369	0.39262	0.60738
	lp–Ge2	(Ge;Pd;3La)	2.0871	0.80815	0.38369	0.61631
(2b)Ge3	Ge3–Ge3	(2Ge;4La)	1.3078	0.48463	1.00000	0.00000
	Ge3–Ge2	(2Ge;4La)	1.5622	0.43848	1.00000	0.00000
	lp–Ge3	(Ge;3La)	2.1002	0.89434	0.21131	0.78869
	lp–Ge3	(Ge;3La)	2.1178	0.89522	0.20956	0.79044
(3b)Ge4	Ge4–Ge4	(2Ge)	2.1330	0.49995	1.00000	0.00000
	Ge4–Ge6	(2Ge;La)	1.9855	0.48829	1.00000	0.00000
	Ge4–Ge6	(2Ge;La)	1.9885	0.48911	1.00000	0.00000
	lp–Ge4	(Ge;Pd;2La)	1.9183	0.70317	0.59365	0.40635
(3b)Ge5	Ge5–Ge5	(2Ge)	2.1260	0.50000	0.99990	0.00010
	Ge5–Ge6	(2Ge;La)	1.9879	0.48740	1.00000	0.00000
	Ge5–Ge6	(2Ge;La)	1.9917	0.48838	1.00000	0.00000
	lp–Ge5	(Ge;Pd;2La)	1.9240	0.70109	0.59782	0.40218
(3b)Ge6	Ge6–Ge6	(2Ge)	2.13840	0.49991	1.00000	0.00000
	Ge6–Ge4	(2Ge;La)	1.9855	0.49373	1.00000	0.00000
	Ge6–Ge5	(2Ge;La)	1.9879	0.49323	1.00000	0.00000
	lp–Ge6	(Ge;Pd;2La)	1.8699	0.72180	0.55639	0.44361

Characteristic quantity referred to each valence basin are listed

Table A4.14 Position-space bonding analysis for Y_2PdGe_6

Central Atom	ELI-D basin (B_i)	Atomicity	Before PSC0 (B_i)	$p(B_i^{Ge})$	cc	lpc	After PSC0 (B_i)	$p(B_i^{Ge})$	cc	lpc
(2b)Ge2	Ge2–Ge2	(2Ge;4Y)	1.6825	0.48392	1.00000	0.00000	1.65317	0.46488	1.00000	0.00000
	Ge2–Ge3	(2Ge;4Y)	1.2997	0.53605	0.92791	0.07209	1.26026	0.52307	0.95387	0.04613
	lp-Ge2	(Ge;Pd;3Y)	2.2477	0.78213	0.43573	0.56427	2.36679	0.70592	0.58816	0.41184
	lp-Ge2	(Ge;Pd;3Y)	2.2478	0.78214	0.43571	0.56429	2.36723	0.70597	0.58805	0.41195
(2b)Ge3	Ge3–Ge3	(2Ge;4Y)	0.7217	0.49674	1.00000	0.00000	0.68022	0.49413	1.00000	0.00000
	Ge3–Ge2	(2Ge;4Y)	1.2997	0.44079	1.00000	0.00000	1.26026	0.42655	1.00000	0.00000
	lp-Ge3	(Ge;3Y)	2.3688	0.89898	0.20204	0.79796	2.49648	0.81110	0.37780	0.62220
	lp-Ge3	(Ge;3Y)	2.3700	0.89886	0.20228	0.79772	2.49748	0.81102	0.37796	0.62204
(3b)Ge4	Ge4–Ge5	(2Ge)	2.0692	0.50058	0.99884	0.00116	1.95255	0.50058	0.99884	0.00116
	Ge4–Ge6	(2Ge;Y)	2.0639	0.48738	1.00000	0.00000	1.99792	0.47496	1.00000	0.00000
	Ge4–Ge6	(2Ge;Y)	2.0654	0.48775	1.00000	0.00000	1.99916	0.47528	1.00000	0.00000
	lp-Ge4	(Ge;Pd;2Y)	1.8986	0.69172	0.61656	0.38344	1.90364	0.65127	0.69745	0.30255
(3b)Ge5	Ge5–Ge4	(2Ge)	2.0692	0.49942	1.00000	0.00000	1.95255	0.49942	1.00000	0.00000
	Ge5–Ge6	(2Ge;Y)	2.0704	0.49106	1.00000	0.00000	2.00519	0.47841	1.00000	0.00000
	Ge5–Ge6	(2Ge;Y)	2.0709	0.49114	1.00000	0.00000	2.00566	0.47848	1.00000	0.00000
	lp-Ge5	(Ge;Pd;2Y)	1.9157	0.68294	0.63413	0.36587	1.93062	0.64014	0.71972	0.28028
(3b)Ge6	Ge6–Ge6	(2Ge)	2.0828	0.50000	1.00000	0.00000	1.96480	0.50000	1.00000	0.00000
	Ge6–Ge4	(2Ge;Y)	2.0654	0.49230	1.00000	0.00000	1.99916	0.47994	1.00000	0.00000
	Ge6–Ge5	(2Ge;Y)	2.0704	0.48846	1.00000	0.00000	2.00519	0.47581	1.00000	0.00000
	lp-Ge6	(Ge;Pd;2Y)	1.8689	0.70180	0.59639	0.40361	1.87275	0.66129	0.67741	0.32259

(continued)

Table A4.14 (continued)

Central Atom	ELI-D basin (B_i)	Atomicity	Before PSC0					After PSC0				
			(B_i)	$p(B_i^{Ge})$ $p(B_i^{Pd})$	cc	lpc		(B_i)	$p(B_i^{Ge})$ $p(B_i^{Pd})$	cc	lpc	
Pd	Pd-Y	(Pd-Y)	0.0718	0.85655	0.28691	0.71309		0.08377	0.72926	0.54148	0.45852	
	Pd-Y	(Pd-Y)	0.0718	0.85655	0.28691	0.71309		0.08377	0.72926	0.54148	0.45852	
	Pd-Y	(Pd-Y)	0.0735	0.85306	0.29388	0.70612		0.08619	0.72259	0.55482	0.44518	
	Pd-Y	(Pd-Y)	0.0735	0.85306	0.29388	0.70612		0.08619	0.72259	0.55482	0.44518	

Characteristic quantity referred to each valence basin, before and after correction PSC0, are listed

Table A5.1 Interatomic distances (<4 Å) in $Nd_2Pd_3Ge_5$

Atom 1	Atom 2	d [Å]	Atom 1	Atom 2	d [Å]
Nd	1Ge2	3.0622(6)	Ge1	4Pd1	2.5268(2)
	1Pd1	3.2104(4)		2Ge1	3.0659(2)
	2Ge1	3.2232(2)		4Nd	3.2232(2)
	1Ge2	3.2725(6)		2Ge3	3.3147(4)
	2Ge2	3.2778(2)	Ge2	1Pd1	2.4263(5)
	2Ge3	3.3148(3)		2Pd2	2.5498(4)
	2Ge3	3.3155(3)		2Ge3	2.6279(4)
	2Pd1	3.3202(2)		1Nd	3.0622(6)
	1Pd1	3.4590(4)		1Nd	3.2725(6)
	2Pd2	3.4837(2)		2Nd	3.2778(2)
	1Pd1	3.4905(4)		2Ge2	3.9962(5)
Pd1	1Ge2	2.4263(5)	Ge3	2Pd1	2.4908(3)
	2Ge3	2.4908(3)		2Ge2	2.6279(4)
	2Ge1	2.5268(2)		1Pd2	2.7124(4)
	1Nd	3.2104(4)		2Ge3	3.0659(2)
	2Nd	3.3202(2)		1Ge1	3.3147(4)
	1Nd	3.4590(4)		2Nd	3.3148(3)
	1Nd	3.4905(4)		2Nd	3.3155(3)
	2Pd1	3.7666(3)			
Pd2	4Ge2	2.5498(4)			
	2Ge3	2.7124(4)			
	2Pd2	3.0659(2)			
	4Nd	3.4837(2)			

Table A5.2 Interatomic distances (< 4 Å) in $Yb_2Pd_3Ge_5$

Atom 1	Atom 2	d [Å]	Atom 1	Atom 2	d [Å]
Yb	1Ge2	3.0417(3)	Ge1	4Pd1	2.5234(2)
	1Pd1	3.1824(3)		2Ge1	2.9913(1)
	2Ge2	3.2119(9)		4Yb	3.2373(1)
	2Ge1	3.2373(1)		2Ge3	3.3570(2)
	2Pd1	3.2484(7)		4Ge2	3.9989(2)
	1Ge2	3.2851(3)	Ge2	1Pd1	2.4192(3)
	2Ge3	3.3062(1)		2Pd2	2.5353(2)
	2Ge3	3.3202(2)		2Ge3	2.6412(2)
	1Pd1	3.5096(2)		1Yb	3.0417(3)
	2Pd2	3.5112(1)		2Yb	3.2119(9)
	1Pd1	3.5257(3)		1Yb	3.2851(3)
Pd1	1Ge2	2.4192(3)		2Ge2	3.8774(2)
	2Ge3	2.5024(2)		2Ge1	3.9989(2)
	2Ge1	2.5234(2)	Ge3	2Pd1	2.5024(2)
	1Yb	3.1824(3)		2Ge2	2.6412(2)
	2Yb	3.2484(7)		1Pd2	2.6721(2)
	1Yb	3.5096(2)		2Ge3	2.9913(1)
	1Yb	3.5257(3)		2Yb	3.3062(1)
	2Pd1	3.7402(1)		2Yb	3.3202(2)
Pd2	4Ge2	2.5353(2)		1Ge1	3.3570(2)
	2Ge3	2.6721(2)			
	2Pd2	2.9913(1)			
	4Yb	3.5112(1)			

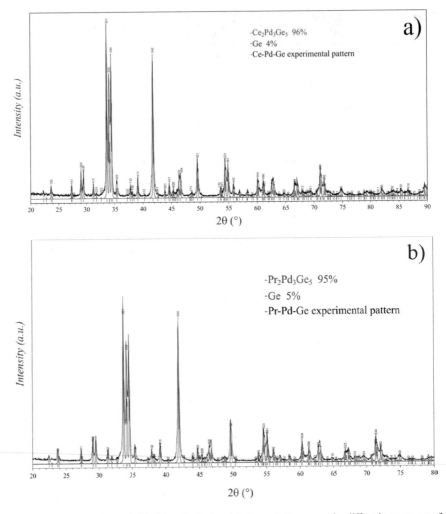

Figure A5.1 Experimental (black) and calculated (coloured) X-ray powder diffraction patterns of selected R-Pd-Ge samples (* indicates unindexed peaks)

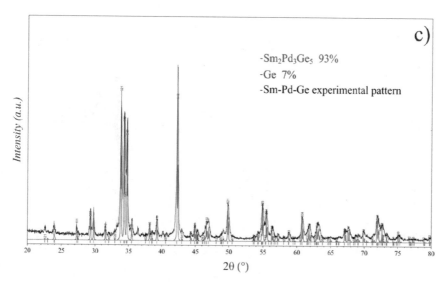

Figure A5.1 (continued)

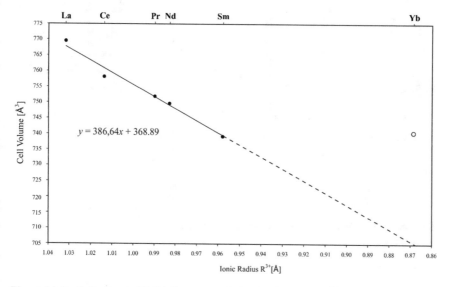

Figure A5.2. Cell volume of $R_2Pd_3Ge_5$ compouds as a function of the R^{3+} ionic radius The empty circle represents the datum for $Yb_2Pd_3Ge_5$ obtained after Rietveld refinement

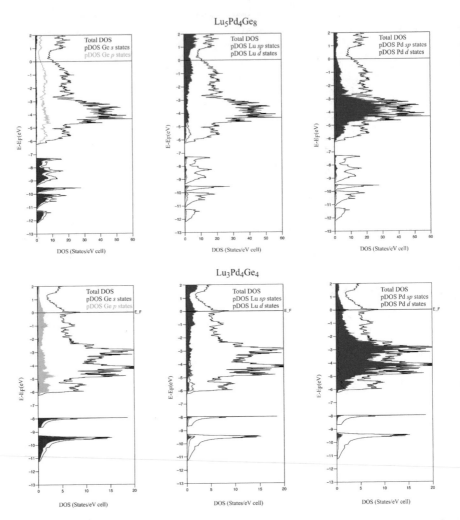

Figure A6.1 Total DOS for $Lu_5Pd_4Ge_8$ and $Lu_3Pd_4Ge_4$ together with the orbital projected DOS for each species

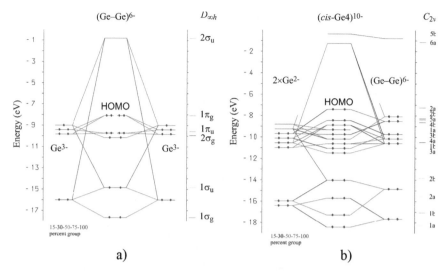

Figure A6.2. Molecular orbitals diagram for Ge_2^{6-} **a** and cis-Ge_4^{10-} **b** as generated by CACAO. For the cis-Ge_4^{10-} unit the point symmetry of the anion was forced to C_{2v} point group fixing all the distances to 2.56 Å and obtuse internal angles to 111°

Printed in the United States
by Baker & Taylor Publisher Services